FREE Study Skills D'

Dear Customer,

Thank you for your purchase from Mometrix! We consider it an honor and privilege that you have purchased our product and want to ensure your satisfaction.

As a way of showing our appreciation and to help us better serve you, we have developed a Study Skills DVD that we would like to give you for FREE. **This DVD covers our "best practices" for studying for your exam, from using our study materials to preparing for the day of the test.**

All that we ask is that you email us your feedback that would describe your experience so far with our product. Good, bad or indifferent, we want to know what you think!

To get your **FREE Study Skills DVD**, email freedvd@mometrix.com with "FREE STUDY SKILLS DVD" in the subject line and the following information in the body of the email:

a. The name of the product you purchased.

b. Your product rating on a scale of 1-5, with 5 being the highest rating.

c. Your feedback. It can be long, short, or anything in-between, just your impressions and experience so far with our product. Good feedback might include how our study material met your needs and will highlight features of the product that you found helpful.

d. Your full name and shipping address where you would like us to send your free DVD.

If you have any questions or concerns, please don't hesitate to contact me directly.

Thanks again!

Sincerely,

Jay Willis
Vice President
jay.willis@mometrix.com
1-800-673-8175

AFOQT SECRETS

Study Guide

Your Key to Exam Success

AFOQT Practice Questions & Review for the Air Force Officer Qualifying Test

Published by

Mometrix Test Preparation

AFOQT Exam Secrets Test Prep Team

Written and edited by the AFOQT Exam Secrets Test Prep Staff

Printed in the United States of America

This paper meets the requirements of ANSI/NISO Z39.48-1992 (Permanence of Paper).

Mometrix offers volume discount pricing to institutions. For more information or a price quote, please contact our sales department at sales@mometrix.com or 888-248-1219.

ISBN 13: 978-1-63094-995-2
ISBN 10: 1-63094-995-7

Dear Future Exam Success Story:

Congratulations on your purchase of our study guide. Our goal in writing our study guide was to cover the content on the test, as well as provide insight into typical test taking mistakes and how to overcome them.

Standardized tests are a key component of being successful, which only increases the importance of doing well in the high-pressure high-stakes environment of test day. How well you do on this test will have a significant impact on your future, and we have the research and practical advice to help you execute on test day.

The product you're reading now is designed to exploit weaknesses in the test itself, and help you avoid the most common errors test takers frequently make.

How to use this study guide

We don't want to waste your time. Our study guide is fast-paced and fluff-free. We suggest going through it a number of times, as repetition is an important part of learning new information and concepts.

First, read through the study guide completely to get a feel for the content and organization. Read the general success strategies first, and then proceed to the content sections. Each tip has been carefully selected for its effectiveness.

Second, read through the study guide again, and take notes in the margins and highlight those sections where you may have a particular weakness.

Finally, bring the manual with you on test day and study it before the exam begins.

Your success is our success

We would be delighted to hear about your success. Send us an email and tell us your story. Thanks for your business and we wish you continued success.

Sincerely,

Mometrix Test Preparation Team

TABLE OF CONTENTS

Top 20 Test Taking Tips

1. Carefully follow all the test registration procedures
2. Know the test directions, duration, topics, question types, how many questions
3. Setup a flexible study schedule at least 3-4 weeks before test day
4. Study during the time of day you are most alert, relaxed, and stress free
5. Maximize your learning style; visual learner use visual study aids, auditory learner use auditory study aids
6. Focus on your weakest knowledge base
7. Find a study partner to review with and help clarify questions
8. Practice, practice, practice
9. Get a good night's sleep; don't try to cram the night before the test
10. Eat a well balanced meal
11. Know the exact physical location of the testing site; drive the route to the site prior to test day
12. Bring a set of ear plugs; the testing center could be noisy
13. Wear comfortable, loose fitting, layered clothing to the testing center; prepare for it to be either cold or hot during the test
14. Bring at least 2 current forms of ID to the testing center
15. Arrive to the test early; be prepared to wait and be patient
16. Eliminate the obviously wrong answer choices, then guess the first remaining choice
17. Pace yourself; don't rush, but keep working and move on if you get stuck
18. Maintain a positive attitude even if the test is going poorly
19. Keep your first answer unless you are positive it is wrong
20. Check your work, don't make a careless mistake

Verbal Analogies

What are analogies questions?

Analogies are pairs of terms that have a common relationship. Analogies questions are presented in the format, "A is to B as C is to D," meaning that terms A and B are related to one another in the same or similar way that terms C and D are related to each other. Terms A and B do not have to be related to terms C and D at all, though they usually will be.

Usually in the question, you will be given terms A, B, and C, and will just have to supply term D from the choices given. Occasionally, you may be given only terms A and B, and you will have to select a pair of terms for C and D.

What sort of relationships will there be?

Below are some examples of the types of analogies that may appear on the exam. Most of the questions you encounter will be relatively simple relationships, but here is an extensive list of the types of analogies that might show up.

Characteristic

Some characteristic analogies will focus on a characteristic of something else.

- Dog : Paw – The foot of a dog is its paw.
- Lady : Lovely – A lady has a lovely personality.
- Outrageous : Lies – Lies can be described as being outrageous.

Some characteristic analogies will focus on something that is NOT a characteristic of something else.

- Desert : Humidity – A desert does not have humidity.
- Job : Unemployed – A person without a job is unemployed.
- Quick : Considered – A quick decision is often not very considered.

Source

- Casting : Metal – A casting is made from metal.
- Forest : Trees – A forest is composed of trees.
- Slogans : Banners – A slogan is printed on banners.

Location

- Eiffel Tower : Paris – The Eiffel Tower is a structure in Paris.
- Welsh : Wales – The Welsh are the inhabitants of Wales.
- Pound : England – The pound is the monetary unit of England.

Sequential

- One : Two – These are consecutive numbers.
- Birth : Death – These are the first and last events of a life or project.
- Spring : Summer – The season of spring immediately precedes summer.

Cause/Effect

- Storm : Hail – Hail can be caused by a storm.
- Heat : Fire – Heat results from a fire.
- Monotony : Boredom – Boredom is a consequence of monotony.

Creator/Creation

- Carpenter : House – A carpenter builds a house.
- Painter : Portrait – A painter makes a portrait.
- Burroughs : Tarzan – Edgar Rice Burroughs wrote the novel Tarzan.

Provider/Provision

- Job : Salary – A job provides a salary.
- Therapist : Treatment – A therapist treats patients.
- Army : Defense – An army enables national defense.

Object/Function

- Pencil : Write – A pencil is used to write.
- Pressure : Barometer – A barometer measures pressure.
- Frown : Unhappy – A frown shows unhappiness.

User/Tool

- Carpenter : Hammer – A carpenter uses a hammer.
- Teacher : Chalk – A teacher uses chalk.
- Farmer : Tractor – A farmer drives a tractor.

Whole/Part

- Door : House – A door is part of a house.
- State : Country – A country is made up of states.
- Day : Month – A month consists of many days.

Grammatical Transformation

- Ran : Run – These are different tenses of the same verb.
- *Die : Dice – These are singular and plural forms.*
- We : Our – These are pronouns related to groups.

Translation

- Satan : Lucifer – These are both names for the devil.
- Bon Voyage : Farewell – These are the French and English words for goodbye.
- Japan : Nippon – These are two names for the same country.

Category

- Door : Window – Both a door and a window are parts of a house.
- Thigh : Shin – Both a thigh and a shin are parts of a leg.
- Measles : Mumps – Both measles and mumps are types of diseases.

Synonym or Definition

These are analogies in which both terms have a similar meaning.

- Chase : Pursue – Both of these terms mean to "go after".
- Achieve : Accomplish – Both of these terms refer to the successful attainment of a goal.
- Satiate : Satisfy – Both of these terms mean to gratify a desire.

Antonym or Contrast

These are analogies in which both terms have an opposite meaning.

- Disguise : Reveal – To disguise something is not to reveal it, but to conceal it.
- Peace : War – Peace is a state in which there is no war.
- Forget : Remember – The word "remember" means not to forget something.

Intensity

These are analogies in which either one term expresses a higher degree of something than the other term.

- Exuberant : Happy – To be exuberant is to be extremely happy.
- Break : Shatter – To shatter is to strongly break.
- Deluge : Rain – A deluge is a heavy rain.

What strategies can I use?

A huge vocabulary is not necessary to succeed on analogies questions (though it certainly doesn't hurt). In most cases though, you can determine the answer even if you don't recognize all the words. The strategies listed here will help you develop the ability to recognize basic relationships and apply simple steps and methods to solving them.

Determine the Relationship

Don't focus on the meanings, but rather the relationship between the two words.
To understand the relationship, first create a sentence that links the two words and puts them into perspective. The sentence that you use to connect the words can be simple at first.

- Example:
  - Wood : Fire
  - *Wood* feeds a *fire*.

Then go through each answer choice and replace the words with the answer choices. If the question is easy, then that may be all that is necessary. If the question is hard, you might have to fine-tune your sentence.

- Example:
  - Fire : Wood :: Cow : (a. grass, *b.* farmer)

Using the initial sentence, you would state "Grass feeds a cow." This is correct, but then so is the next answer choice "Farmer feeds a cow." So which is right? Modify the sentence to be more specific.

- Example: "Wood feeds a fire and is consumed."

This modified sentence makes answer choice B incorrect and answer choice A clearly correct, because while "Grass feeds a cow and is consumed" is correct, "farmer feeds a cow and is consumed" is definitely wrong.

If your initial sentence seems correct with more than one answer choice, then keep modifying it until only one answer choice makes sense.

Similar Choices

If you don't know the word, don't worry. Look at the answer choices and just use them. Remember that three of the answer choices will always be wrong. If you can find a common relationship between any three answer choices, then you know they are all wrong. Find the answer choice that does not have a common relationship to the other answer choices and it will be the correct answer.

- Example:
  - Tough : Rugged :: Hard : (a. soft, *b.* easy, *c.* delicate, *d.* rigid)

In this example the first three choices are all opposites of the term "hard". Even if you don't know that rigid means the same as hard, you know it must be correct, because the other three all had the same relationship. They were all opposites, so they must all be wrong. The one that has a different relationship from the other three must be correct. So don't worry if you don't know a word. Focus on the answer choices that you do understand and see if you can identify common relationships. Even identifying two word pairs with the same relationship (for example, two word pairs that are both opposites) will allow you to eliminate those two answer choices, for they are both wrong.

A simple way to remember this is that if you have two or more answer choices that have the exact same relationship, then they are both or all wrong.

- Example: (a. neat, *b.* orderly)

Since the two answer choices above are synonyms and therefore have the same relationship with the matching term, then you know that they both must be wrong, because they both can't be correct, and for all intents and purposes they are the same word.

Be sure to read all of the choices. You may find an answer choice that seems right at first, but continue reading and you may find a better choice.

Difficult words are usually synonyms or antonyms (opposites). Whenever you have extremely difficult words that you don't understand, look at the answer choices. Try and identify whether two or more of the answer choices are either synonyms or antonyms. Remember that if you can find two word pairs that have the same relationship (for example, they are both synonyms) then you can eliminate them both.

Eliminate Answers

Eliminate choices as soon as you realize they are wrong, but be careful! Make sure you consider all of the possible answer choices. Don't worry if you are stuck between two that seem right. By eliminating the other two possible choices, your odds are now 50/50. Rather than wasting too much time, play the odds. You are guessing, but guessing wisely, because you've been able to knock out some of the answer choices that you know are wrong. If you are eliminating choices and realize that the answer choice you are left with is also obviously wrong, don't panic. Start over and consider each choice again. There may easily be something that you missed the first time and will realize on the second pass.

Word Types

The correct answer choice will contain words that are the same type of word as those in the word pair.

- Example:
  - Artist : Paintbrush

In this example, an artist is a person, while a paintbrush is an object. The correct answer will have one word that describes a person and another word that describes an object.

- Example:
  - Hedge : Gardener :: Rock : (a. wind, *b.* sculptor)

In this example, you could create the sentence, "Gardener cuts away at hedges." Both answer choices seem correct with this sentence, "Wind cuts away at rocks" through the process of erosion, and "sculptor cuts away at rocks" using a hammer and chisel. The difference is that a gardener is a person, as is a sculptor, while the wind is a thing, which makes answer choice B correct.

Nearly and Perfect Opposites

When you have determined which pair of terms you should work with, and know that the provided pair is an opposite, then you must find the opposite of the remaining unmatched term. Nearly opposite may often be more correct, because the goal is to test your understanding of the nuances, or little differences, between words. A perfect opposite may not exist, so don't be concerned if your answer choice is not a complete opposite. Focus upon edging closer to the word. Eliminate the words that you know aren't correct first. Then narrow your search. Cross out the words that are the most similar to the main word until you are left with the one that is the least similar.

Prefixes

Take advantage of every clue that the word might include. Prefixes and suffixes can be a huge help. Usually they allow you to determine a basic meaning. Pre- means before, post- means after, pro – is positive, de- is negative. From these prefixes and suffixes, you can get an idea of the general meaning of the word and look for its opposite. Beware though of any traps. Just because con is the opposite of pro, doesn't necessarily mean congress is the opposite of progress!

Positive vs. Negative

Many words can be easily determined to be a positive word or a negative word. Words such as despicable, and gruesome, bleak are all negative. Words such as ecstatic, praiseworthy, and magnificent are all positive. You will be surprised at how many words can be considered as either positive or negative. If you recognize a positive/negative relationship between the given pair of terms, then focus in on the answer choices that would duplicate that positive/negative relationship with the remaining term.

Word Strength

When analyzing a word, determine how strong it is. For example, stupendous and good are both positive words. However, stupendous is a much stronger positive adjective than good. Also, towering or gigantic are stronger words than tall or large. Search for an answer choice with either the same or opposite strength (depending on the relationship of the matched terms) to the remaining term.

Type and Topic

Another key is what type of word is the unmatched term. If the unmatched term is an adjective describing height, then look for the answer choice to be an adjective describing height as well. Match both the type and topic of the main word. The type refers the parts of speech, whether the word is an adjective, adverb, or verb. The topic refers to what the definition of the word includes, such as descriptive sizes (large, small, gigantic, etc).

Form a Sentence

Many words seem more natural in a sentence. *Specious* reasoning, *irresistible* force, and *uncanny* resemblance are just a few of the word combinations that usually go together. When faced with an uncommon word that you barely understand, try to put the word in a sentence that makes sense. It will help you to understand the word's meaning and make it easier to determine its relationship. Once you have a good descriptive sentence that utilizes a main term or answer choice properly, plug in the answer choice or main term and see if a solid relationship can be established.

Use Logic

Ask yourself questions about each answer choice to see if they are logical.

- Example:
  - Aromas : Smelt :: Poundings : (a. seen, *b.* heard)

Would poundings be "seen"? Or would pounding be "heard"? It can logically be deduced that poundings are heard.

The Trap of Familiarity

Don't just choose a word because you recognize it. On difficult questions, you may only recognize one or two words. There won't be any made up words on the test, so don't think that just because you only recognize one word means that word must be correct. If you don't recognize three words, then focus on the one that you do recognize. Is it correct? Try your best to determine if it fits the sentence you've created that shows the relationship between terms. If it does, that is great, but if it doesn't, eliminate it. Each word you eliminate increases your chances of getting the question correct.

Tough Questions

If you are stumped on a problem or it appears too hard or too difficult, don't waste time. Move on! Remember though, if you can quickly check for obviously incorrect answer choices, your chances of guessing correctly are greatly improved. Before you completely give up, at least try to knock out a couple of possible answers. Eliminate what you can and then guess at the remainder before moving on.

Read Carefully

Understand the analogy. Read the terms and answer choices carefully. Don't miss the question because you misread the terms. There are only a few words in each question, so you can spend time reading them carefully. Yet a happy medium must be attained, so don't waste too much time. You must read carefully, but efficiently.

Brainstorm

If you get stuck on a difficult analogy, spend a few seconds quickly brainstorming. Run through the complete list of possible relationships. Break down each answer choice into all of the potential combinations with the two possible analogous terms. Since there are four answer choices and each answer choice could form a pair with one of two terms, then there are only eight possible relationships to test. Look at each relationship and see if it would make sense. Test with sentences to determine if any relationship can be established. By systematically going through all possibilities, you may find something that you would otherwise overlook.

Practice Questions

1. STIRRUP is to EAR as ATRIUM is to
    a. blood
    b. ventricle
    c. vestibule
    d. heart
    e. chamber

2. DWELLING is to CONDOMINIUM as MEAL is to
    a. entree
    b. brunch
    c. appetizer
    d. plate
    e. dessert

3. FRAGRANT is to SMELL as MELLIFLUOUS is to
    a. sound
    b. pleasant
    c. fluid
    d. taste
    e. soothing

4. BRACELET is to JEWELRY as POMEGRANATE is to
    a. seeds
    b. edible
    c. fruit
    d. acidic
    e. citrus

5. LOUD is to DEAFENING as HAPPY is to
    a. ecstatic
    b. glad
    c. morose
    d. content
    e. depressed

6. FACILE is to EASY as LOQUACIOUS is to
    a. silent
    b. difficult
    c. friendly
    d. lacking
    e. talkative

7. ABATE is to INCREASE as ABHOR
   a. despise
   b. love
   c. tolerate
   d. abdicate
   e. hate

8. QUILLS is to PORCUPINE as TUSKS is to
   a. proboscis
   b. horn
   c. ivory
   d. elephant
   e. nose

9. DUPLICITY is to DECEPTION as AVARICE is to
   a. greed
   b. money
   c. average
   d. accumulate
   e. benevolent

10. WHISPER is to YELL as TAP is to
   a. water
   b. pat
   c. dance
   d. jab
   e. jump

11. EUCALYPTUS is to TREE as IRIS is to
   a. tulip
   b. purple
   c. eye
   d. face
   e. flower

12. IRKSOME is to TEDIOUS as INTRIGUING is to
   a. fascinating
   b. silly
   c. unlikely
   d. impossible
   e. irritating

Practice Answers

1. D: This is a "part to whole" analogy. Just as the stirrup is a part of the ear, so is the atrium a part of the heart.

2. B: In "type" analogies, one word in the stem names a category that encompasses the other. Just as a condominium is a type of dwelling, so is brunch a type of meal.

3. A: This analogy is that of adjective to noun. *Fragrant* is an adjective modifying the noun *smell* in a positive way. *Mellifluous* is an adjective modifying the noun *sound* in a positive way.

4. C: In this "type" analogy, one word in the stem names a category that encompasses the other. A bracelet is a type of jewelry, just as a pomegranate is a type of fruit.

5. A: In this analogy of relative degree, the second term in each pair indicates a more intense degree of the first term. *Deafening* is a more intense version of *loud,* just as *ecstatic* is a more intense version of *happy.*

6. E: This analogy is based on synonyms. Just as *facile* and *easy* mean about the same thing, so do *loquacious* and *talkative.*

7. B: This analogy is based on antonyms. Just as *abate* means the opposite of *increase,* so does *abhor* mean the opposite of *love.*

8. D: This analogy names prominent features of each animal. Quills are a prominent feature of a porcupine, just as tusks are a prominent feature of an elephant.

9. A: This is another synonym-based analogy. *Duplicity* and *deception* mean about the same thing, just as *avarice* and *greed* do. You might have been tempted to choose D, which suggests something that an avaricious person might do, but *greed* is the better answer since it is a noun, as *avarice* is.

10. D: This is an analogy of relative degree. A yell is a much louder version of a whisper, just as a jab is a much harder version of a tap.

11. E: This is an analogy indicating types, since a eucalyptus is one type of tree, and an iris is one type of flower.

12. A: In this analogy based on synonyms, *irksome* means about the same as *tedious,* just as *intriguing* means about the same as *fascinating.*

Arithmetic Reasoning

What do arithmetic questions look like?

Arithmetic questions will generally take the form of a simple word problem. You will be posed an everyday situation that requires arithmetic to solve and asked to select the correct answer from the choices given. You may be asked to calculate rates, percentages, averages, or some other practical math quantity, and you may have to convert between different units. It is usually obvious what the question is asking for.

How can I prepare?

Since the math needed for these questions is not complicated, that means that you only need to learn or refresh your memory of a few simple operations. Then it's just a matter of practicing them. One of the biggest mistakes people make when trying to learn math is they read about a concept, look at a worked out example problem, and when it makes sense, they assume they understand it well enough and move on. Then when the test comes, they don't remember how to solve the problems. Math skills must be practiced in order to be remembered.

What math skills do I need?

Below are all of the major concepts needed to excel on the arithmetic section:

Operations

There are four basic mathematical operations:
Addition increases the value of one quantity by the value of another quantity. Example: $2 + 4 = 6; 8 + 9 = 17$. The result is called the sum. With addition, the order does not matter. $4 + 2 = 2 + 4$.
Subtraction is the opposite operation to addition; it decreases the value of one quantity by the value of another quantity. Example: $6 - 4 = 2; 17 - 8 = 9$. The result is called the difference. Note that with subtraction, the order does matter. $6 - 4 \neq 4 - 6$.
Multiplication can be thought of as repeated addition. One number tells how many times to add the other number to itself. Example: 3×2 (three times two) $= 2 + 2 + 2 = 6$. With multiplication, the order does not matter. $2 \times 3 = 3 \times 2$ or $3 + 3 = 2 + 2 + 2$.
Division is the opposite operation to multiplication; one number tells us how many parts to divide the other number into. Example: $20 \div 4 = 5$; if 20 is split into 4 equal parts, each part is 5. With division, the order of the numbers does matter. $20 \div 4 \neq 4 \div 20$.

Fractions, percentages, and related concepts

A fraction is a number that is expressed as one integer written above another integer, with a dividing line between them $\left(\frac{x}{y}\right)$. It represents the quotient of the two numbers "x divided by y." It can also be thought of as x out of y equal parts.

The top number of a fraction is called the numerator, and it represents the number of parts under consideration. The 1 in $\frac{1}{4}$ means that 1 part out of the whole is being considered in the calculation. The bottom number of a fraction is called the denominator, and it represents the total number of equal parts. The 4 in $\frac{1}{4}$ means that the whole consists of 4 equal parts. A fraction cannot have a denominator of zero; this is referred to as "undefined."
Fractions can be manipulated, without changing the value of the fraction, by multiplying or dividing (but not adding or subtracting) both the numerator and denominator by the same number. If you divide both numbers by a common factor, you are reducing or simplifying the fraction. Two fractions that have the same value, but are expressed differently are known as equivalent fractions. For example, $\frac{2}{10}, \frac{3}{15}, \frac{4}{20}$, and $\frac{5}{25}$ are all equivalent fractions. They can also all be reduced or simplified to $\frac{1}{5}$.

When two fractions are manipulated so that they have the same denominator, this is known as finding a common denominator. The number chosen to be that common denominator should be the least common multiple of the two original denominators. Example: $\frac{3}{4}$ and $\frac{5}{6}$; the least common multiple of 4 and 6 is 12. Manipulating to achieve the common denominator: $\frac{3}{4} = \frac{9}{12}; \frac{5}{6} = \frac{10}{12}$.

If two fractions have a common denominator, they can be added or subtracted simply by adding or subtracting the two numerators and retaining the same denominator. Example: $\frac{1}{2} + \frac{1}{4} = \frac{2}{4} + \frac{1}{4} = \frac{3}{4}$.
If the two fractions do not already have the same denominator, one or both of them must be manipulated to achieve a common denominator before they can be added or subtracted.

Two fractions can be multiplied by multiplying the two numerators to find the new numerator and the two denominators to find the new denominator. Example: $\frac{1}{3} \times \frac{2}{3} = \frac{1 \times 2}{3 \times 3} = \frac{2}{9}$.
Two fractions can be divided flipping the numerator and denominator of the second fraction and then proceeding as though it were a multiplication. Example: $\frac{2}{3} \div \frac{3}{4} = \frac{2}{3} \times \frac{4}{3} = \frac{8}{9}$.

A fraction whose denominator is greater than its numerator is known as a proper fraction, while a fraction whose numerator is greater than its denominator is known as an improper fraction. Proper fractions have values less than one and improper fractions have values greater than one.

A mixed number is a number that contains both an integer and a fraction. Any improper fraction can be rewritten as a mixed number. Example: $\frac{8}{3} = \frac{6}{3} + \frac{2}{3} = 2 + \frac{2}{3} = 2\frac{2}{3}$. Similarly, any mixed number can be rewritten as an improper fraction. Example: $1\frac{3}{5} = 1 + \frac{3}{5} = \frac{5}{5} + \frac{3}{5} = \frac{8}{5}$.

Percentages can be thought of as fractions that are based on a whole of 100; that is, one whole is equal to 100%. The word percent means "per hundred." Fractions can be expressed as percents by finding equivalent fractions with a denomination of 100. Example: $\frac{7}{10} = \frac{70}{100} = 70\%; \frac{1}{4} = \frac{25}{100} = 25\%$.
To express a percentage as a fraction, divide the percentage number by 100 and reduce the fraction to its simplest possible terms. Example: $60\% = \frac{60}{100} = \frac{3}{5}; 96\% = \frac{96}{100} = \frac{24}{25}$.
Converting decimals to percentages and percentages to decimals is as simple as moving the decimal point. To convert from a decimal to a percent, move the decimal point two places to the right. To convert from a percent to a decimal, move it two places to the left. Example: 0.23 = 23%; 5.34 = 534%; 0.007 = 0.7%; 700% = 7.00; 86% = 0.86; 0.15% = 0.0015.

It may be helpful to remember that the percentage number will always be larger than the equivalent decimal number.

A percentage problem can be presented three main ways: (1) Find what percentage of some number another number is. Example: What percentage of 40 is 8? (2) Find what number is some percentage of a given number. Example: What number is 20% of 40? (3) Find what number another number is a given percentage of. Example: What number is 8 20% of? The three components in all of these cases are the same: a whole (W), a part (P), and a percentage (%). These are related by the equation: $P = W \times \%$. This is the form of the equation you would use to solve problems of type (2). To solve types (1) and (3), you would use these two forms: $\% = \frac{P}{W}$ and $W = \frac{P}{\%}$.

The thing that frequently makes percentage problems difficult is that they are most often also word problems, so a large part of solving them is figuring out which quantities are what. Example: In a school cafeteria, 7 students choose pizza, 9 choose hamburgers, and 4 choose tacos. Find the percentage that chooses tacos. To find the whole, you must first add all of the parts: 7 + 9 + 4 = 20. The percentage can then be found by dividing the part by the whole ($\% = \frac{P}{W}$): $\frac{4}{20} = \frac{20}{100} = 20\%$.

A ratio is a comparison of two quantities in a particular order. Example: If there are 14 computers in a lab, and the class has 20 students, there is a student to computer ratio of 20 to 14, commonly written as 20:14. Ratios are normally reduced to their smallest whole number representation, so 20:14 would be reduced to 10:7 by dividing both sides by 2.

A proportion is a relationship between two quantities that dictates how one changes when the other changes. A direct proportion describes a relationship in which a quantity increases by a set amount for every increase in the other quantity, or decreases by that same amount for every decrease in the other quantity. Example: Assuming a constant driving speed, the time required for a car trip increases as the distance of the trip increases. The distance to be traveled and the time required to travel are directly proportional.

Inverse proportion is a relationship in which an increase in one quantity is accompanied by a decrease in the other, or vice versa. Example: the time required for a car trip decreases as the speed increases, and increases as the speed decreases, so the time required is inversely proportional to the speed of the car.

Area formulas

Rectangle: $A = wl$, where w is the width and l is the length

Square: $A = s^2$, where s is the length of a side.

Triangle: $A = \frac{1}{2}bh$, where b is the length of one side (base) and h is the distance from that side to the opposite vertex measured perpendicularly (height).

Circle: $A = \pi r^2$, where π is the mathematical constant approximately equal to 3.14 and r is the distance from the center of the circle to any point on the circle (radius).

Volume formulas

Rectangular Prism – all 6 sides are rectangles. The volume can be calculated as $V = s_1 \times s_2 \times s_3$, or the lengths of the three different sides multiplied together.

Cube – a special type of prism in which all faces are squares. The volume can be calculated as $V = s^3$, where s is the length of any side.

Sphere – a round solid consisting of one continuous, uniformly-curved surface. The volume can be calculated as $V = \frac{4}{3}\pi r^3$, where r is the distance from the center of the sphere to any point on the surface (radius).

Practice Questions

1. A couple plans to buy a car. They have $569 in a joint bank account. The man has $293 in additional cash and the woman has $189. What is the most expensive down payment they will be able to afford?

   a. $482
   b. $758
   c. $862
   d. $1051
   e. $1121

2. The temperature of a cup of coffee is 98 degrees. If its temperature decreases by 2 degrees per minute, what will its temperature be after 4 minutes?

   a. 100 degrees
   b. 98 degrees
   c. 94 degrees
   d. 90 degrees
   e. 88 degrees

3. A man's lawn grass is 3 inches high. He mows the lawn and cuts off 30% of its height. How tall will the grass be after the lawn is mowed?

   a. 0.9 inches
   b. 2.1 inches
   c. 2.7 inches
   d. 2.9 inches
   e. 3.3 inches

4. Three outlets are selling concert tickets. One ticket outlet sells 432; another outlet sells 238; the third outlet sells 123. How many concert tickets were sold in total?

   a. 361
   b. 555
   c. 670
   d. 793
   e. 823

5. A boy has a bag with 26 pieces of candy inside. He eats 8 pieces of candy, then divides the rest evenly between two friends. How many pieces of candy will each friend get?

   a. 7
   b. 9
   c. 11
   d. 13
   e. 18

Practice Answers

1. D: Calculate the total amount of money the couple has available to spend, which is the amount in the joint bank account and the amount that each has:
$569 + $293 + $189 = $1051

2. D: First, find out what the total temperature decrease will be after 4 minutes:
2 * 4 = 8 degrees
Then, subtract that from the original temperature: 98 – 8 = 90 degrees

3. B: First, calculate 30% of 3 inches: 3 * 0.3 = 0.9 inches.
Then, subtract this value from the original length: 3 – 0.9 = 2.1

4. D: Add the number of tickets that were sold at each location to get the total number of tickets sold: 432 + 238 + 123 = 793

5. B: First, figure out how many pieces of candy are in the bag before they are divided:
26 – 8 = 18
Then, figure out how many pieces each friend will get by dividing by 2: 18 / 2 = 9

Word Knowledge

What do word knowledge questions look like?

Word knowledge questions follow a very simple format. You will be given a word and you must select the word that is closest in meaning to the given word, from the choices given.

How can I prepare?

Unfortunately, there is no easy way to prepare for this section. The questions test your vocabulary and if you don't have a very large vocabulary, this will probably be one of your harder sections to prepare for. The best way to build a large, long-lasting vocabulary is to read extensively, but if your test is approaching, odds are you don't have time for that. Cramming with vocabulary lists is one way to build up a short term vocabulary. Learning common prefixes and suffixes is another valuable use of limited time. Whether or not you decide you need to add to your vocabulary before the test, there are a few strategies you can use to get the most out of the words you already know.

What strategies can I use?

Nearly and Perfect Synonyms
You must determine which of the provided choices has the best similar definition as a certain word. Nearly similar may often be more correct, because the goal is to test your understanding of the nuances, or little differences, between words. A perfect match may not exist, so don't be concerned if your answer choice is not a complete synonym. Focus upon edging closer to the word. Eliminate the words that you know aren't correct first. Then narrow your search. Cross out the words that are the least similar to the main word until you are left with the one that is the most similar.

Prefixes
Take advantage of every clue that the word might include. Prefixes and suffixes can be a huge help. Usually they allow you to determine a basic meaning. Pre- means before, post- means after, pro – is positive, de- is negative. From these prefixes and suffixes, you can get an idea of the general meaning of the word and look for its opposite. Beware though of any traps. Just because con is the opposite of pro, doesn't necessarily mean congress is the opposite of progress!

Positive vs. Negative
Many words can be easily determined to be a positive word or a negative word. Words such as despicable, gruesome, and bleak are all negative. Words such as ecstatic, praiseworthy, and magnificent are all positive. You will be surprised at how many words can be considered as either positive or negative. Once that is determined, you can quickly eliminate any other words with an opposite meaning and focus on those that have the other characteristic, whether positive or negative.

Word Strength
Part of the challenge is determining the most nearly similar word. This is particularly true when two words seem to be similar. When analyzing a word, determine how strong it is. For example, stupendous and good are both positive words.

However, stupendous is a much stronger positive adjective than good. Also, towering or gigantic are stronger words than tall or large. Search for an answer choice that is similar and also has the same strength. If the main word is weak, look for similar words that are also weak. If the main word is strong, look for similar words that are also strong.

Type and Topic

Another key is what type of word is the main word. If the main word is an adjective describing height, then look for the answer to be an adjective describing height as well. Match both the type and topic of the main word. The type refers the parts of speech, whether the word is an adjective, adverb, or verb. The topic refers to what the definition of the word includes, such as sizes or fashion styles.

Form a Sentence

Many words seem more natural in a sentence. *Specious* reasoning, *irresistible* force, and *uncanny* resemblance are just a few of the word combinations that usually go together. When faced with an uncommon word that you barely understand, try to put the word in a sentence that makes sense. It will help you to understand the word's meaning. Once you have a good descriptive sentence that utilizes the main word properly, plug in the answer choices and see if the sentence still has the same meaning with each answer choice. The answer choice that maintains the meaning of the sentence is correct!

Use Replacements

Using a sentence is a great help because it puts the word into a proper perspective. Since the exam actually gives you a sentence, sometimes you don't always have to create your own (though in many cases the sentence won't be helpful). Read the provided sentence, picking out the main word. Then read the sentence again and again, each time replacing the main word with one of the answer choices. The correct answer should "sound" right and fit.
Example: The desert landscape was desolate. Desolate means

a. cheerful
b. creepy
c. excited
d. forlorn

After reading the example sentence, begin replacing "desolate" with each of the answer choices. Does "the desert landscape was cheerful, creepy, excited, or forlorn" sound right? Deserts are typically hot, empty, and rugged environments, probably not cheerful, or excited. While creepy might sound right, that word would certainly be more appropriate for a haunted house. But "the desert landscape was forlorn" has a certain ring to it and would be correct.

Eliminate Similar Choices

If you don't know the word, don't worry. Look at the answer choices and just use them. Remember that three of the answer choices will always be wrong. If you can find a common relationship between any three answer choices, then you know they are wrong. Find the answer choice that does not have a common relationship to the other answer choices and it will be the correct answer.
Example: Laconic most nearly means

a. wordy
b. talkative
c. expressive
d. quiet

In this example, the first three choices are all similar. Even if you don't know that laconic means the same as quiet, you know that "quiet" must be correct, because the other three choices were all virtually the same. They were all the same, so they must all be wrong. The one that is different must be correct. So, don't worry if you don't know a word. Focus on the answer choices that you do understand and see if you can identify similarities. Even identifying two words that are similar will allow you to eliminate those two answer choices. Because they are similar, they are either both right or both wrong, and since they can't both be right, they must both be wrong.
Example: He worked slowly, moving the leather back and forth until it was ____.

a. rough
b. hard
c. stiff
d. pliable

In this example the first three choices are all similar and synonyms. Even without knowing what pliable means, it has to be correct, because you know the other three answer choices mean the same thing.

Adjectives Give it Away

Words mean things and are added to the sentence for a reason. Adjectives in particular may be the clue to determining which answer choice is correct.
Example: The brilliant scientist made several discoveries that were

a. dull
b. dazzling

Look at the adjectives first to help determine what makes sense. A "brilliant" or smart scientist would make dazzling, rather than dull discoveries. Without that simple adjective, no answer choice is clear.

Use Logic

Ask yourself questions about each answer choice to see if they are logical.
Example: In the distance, the deep pounding resonance of the drums could be

a. seen
b. heard

Would resonating pounding be seen or would resonating pounding be heard?

The Trap of Familiarity

Don't just choose a word because you recognize it. On difficult questions, you may only recognize one or two words. The exam doesn't have "make-believe words" on it, so don't think that just because you only recognize one word means that word must be correct. If you don't recognize most of the words, then focus on the ones that you do recognize. Are any of them correct? Try your best to determine if they fit the sentence. If any of them do, you have your answer, but if not, eliminate them and guess from among the remaining options.

Practice Questions

1. **Sketch** most nearly means
    a. skip
    b. scope
    c. draw
    d. drain
    e. drip

2. The child was **frightened** by the movie.
    a. scared
    b. entertained
    c. amused
    d. saddened
    e. delighted

3. **Sever** most nearly means
    a. hard
    b. cut
    c. add
    d. soft
    e. change

4. Her prediction was **accurate**.
    a. false
    b. funny
    c. planned
    d. assumed
    e. correct

5. **Taunt** most nearly means
    a. truant
    b. tried
    c. tight
    d. tease
    e. tired

6. Her concern for him was **sincere.**
    a. intense
    b. genuine
    c. brief
    d. misunderstood
    e. repetitive

7. **Disclose** most nearly means
a. reveal
b. return
c. near
d. hide
e. conceal

8. He **sprinted** down the road.
a. crawled
b. walked
c. hurried
d. ran
e. drove

9. The naughty child was **disciplined**.
a. upset
b. punished
c. hidden
d. bad
e. surprised

10. **Seize** most nearly means
a. grab
b. release
c. tell
d. fight
e. give

Practice Answers

1. C: To sketch something is to draw something. Saying somebody was planning to sketch a landscape and saying they were going to draw a landscape conveys the same meaning.

2. A: To say somebody is frightened is the same as saying they are scared or afraid.

3. B: To sever something is to cut something. For example, to say that somebody severed all ties with somebody else means that they have cut those ties. It can also be used to describe the cutting of objects. For example, saying someone severed a rope with a knife means they cut the rope.

4. E: Describing something as accurate and describing it as correct conveys the same meaning. For example, saying somebody accurately predicted something is the same as saying they correctly predicted something.

5. D: To taunt somebody is to tease them. To say somebody taunted another person conveys the same meaning as saying somebody teased another person. Usually, teasing and taunting is understood to be a mean practice.

6. B: To say something is sincere means that it is genuine or real. For example, saying someone showed sincere concern means that their concern was genuine, and not fake.

7. A: To disclose something is to reveal something. For example, saying somebody disclosed something they had been hiding is the same as saying they revealed it.

8. D: To sprint is to run. Saying that somebody sprinted to their destination and saying they ran to their destination conveys the same meaning.

9. B: To discipline someone for their undesirable actions or behaviors is to punish them. Saying a child was disciplined for his actions and saying he was punished conveys the same meaning.

10. A: To seize something is to take hold of it or grab it. For example, saying the woman seized the man's arm is the same as saying she grabbed it.

Math Knowledge

What do math knowledge questions look like?

Math knowledge questions are much less predictable than arithmetic questions. Math knowledge questions may test your knowledge of anything covered in the arithmetic section, plus square roots, exponents, factors, multiples, equations, geometric properties, and more.

How can I prepare?

The questions may be more difficult, but the preparation process should be the same: learn the concepts and facts you need to know, and then practice them.

What math skills do I need?

In addition to everything covered in the arithmetic section, you may also need to know the following concepts:

Numbers and their classifications

Numbers are the basic building blocks of mathematics. Specific features of numbers are identified by the following terms:
Integers – The set of whole positive and negative numbers, including zero. Integers do not include fractions $\left(\frac{1}{3}\right)$, decimals (0.56), or mixed numbers $\left(7\frac{3}{4}\right)$.
Prime number – A whole number greater than 1 that has only two factors, itself and 1; that is, a number that can be divided evenly only by 1 and itself.
Composite number – A whole number greater than 1 that has more than two different factors; in other words, any whole number that is not a prime number. For example: The composite number 8 has the factors of 1, 2, 4, and 8.
Even number – Any integer that can be divided by 2 without leaving a remainder. For example: 2, 4, 6, 8, and so on.
Odd number – Any integer that cannot be divided evenly by 2. For example: 3, 5, 7, 9, and so on.
Decimal number – a number that uses a decimal point to show the part of the number that is less than one. Example: 1.234.
Decimal point – a symbol used to separate the ones place from the tenths place in decimals or dollars from cents in currency.
Decimal place – the position of a number to the right of the decimal point. In the decimal 0.123, the 1 is in the first place to the right of the decimal point, indicating tenths; the 2 is in the second place, indicating hundredths; and the 3 is in the third place, indicating thousandths.

The decimal, or base 10, system is a number system that uses ten different digits (0, 1, 2, 3, 4, 5, 6, 7, 8, 9). An example of a number system that uses something other than ten digits is the binary, or base 2, number system, used by computers, which uses only the numbers 0 and 1. It is thought that the decimal system originated because people had only their 10 fingers for counting.

Rational, irrational, and real numbers can be described as follows:
Rational numbers include all integers, decimals, and fractions. Any terminating or repeating decimal number is a rational number.
Irrational numbers cannot be written as fractions or decimals because the number of decimal places is infinite and there is no recurring pattern of digits within the number. For example, pi (π) begins with 3.141592 and continues without terminating or repeating, so pi is an irrational number.
Real numbers are the set of all rational and irrational numbers.

Operations

An exponent is a superscript number placed next to another number at the top right. It indicates how many times the base number is to be multiplied by itself. Exponents provide a shorthand way to write what would be a longer mathematical expression. Example: $a^2 = a \times a$; $2^4 = 2 \times 2 \times 2 \times 2$. A number with an exponent of 2 is said to be "squared," while a number with an exponent of 3 is said to be "cubed." The value of a number raised to an exponent is called its power. So, 8^4 is read as "8 to the 4th power," or "8 raised to the power of 4." A negative exponent is the same as the reciprocal of a positive exponent. Example: $a^{-2} = \frac{1}{a^2}$.

Parentheses are used to designate which operations should be done first when there are multiple operations. Example: $4 - (2 + 1) = 1$; the parentheses tell us that we must add 2 and 1, and then subtract the sum from 4, rather than subtracting 2 from 4 and then adding 1 (this would give us an answer of 3).

Order of Operations is a set of rules that dictates the order in which we must perform each operation in an expression so that we will evaluate at accurately. If we have an expression that includes multiple different operations, Order of Operations tells us which operations to do first. The most common mnemonic for Order of Operations is PEMDAS, or "Please Excuse My Dear Aunt Sally." PEMDAS stands for Parentheses, Exponents, Multiplication, Division, Addition, Subtraction. It is important to understand that multiplication and division have equal precedence, as do addition and subtraction, so those pairs of operations are simply worked from left to right in order.
Example: Evaluate the expression $5 + 20 \div 4 \times (2 + 3)^2 - 6$ using the correct order of operations.
P: Perform the operations inside the parentheses, $(2 + 3) = 5$.
E: Simplify the exponents, $(5)^2 = 25$.
The equation now looks like this: $5 + 20 \div 4 \times 25 - 6$.
MD: Perform multiplication and division from left to right, $20 \div 4 = 5$; then $5 \times 25 = 125$.
The equation now looks like this: $5 + 125 - 6$.
AS: Perform addition and subtraction from left to right, $5 + 125 = 130$; then $130 - 6 = 124$.

The laws of exponents are as follows:
1) Any number to the power of 1 is equal to itself: $a^1 = a$.
2) The number 1 raised to any power is equal to 1: $1^n = 1$.
3) Any number raised to the power of 0 is equal to 1: $a^0 = 1$.
4) Add exponents to multiply powers of the same base number: $a^n \times a^m = a^{n+m}$.
5) Subtract exponents to divide powers of the same number; that is $a^n \div a^m = a^{n-m}$.
6) Multiply exponents to raise a power to a power: $(a^n)^m = a^{n \times m}$.
7) If multiplied or divided numbers inside parentheses are collectively raised to a power, this is the same as each individual term being raised to that power: $(a \times b)^n = a^n \times b^n$; $(a \div b)^n = a^n \div b^n$.

Note: Exponents do not have to be integers. Fractional or decimal exponents follow all the rules above as well. Example: $5^{\frac{1}{4}} \times 5^{\frac{3}{4}} = 5^{\frac{1}{4}+\frac{3}{4}} = 5^1 = 5$.
A root, such as a square root, is another way of writing a fractional exponent. Instead of using a superscript, roots use the radical symbol ($\sqrt{\ }$) to indicate the operation. A radical will have a number underneath the bar, and may sometimes have a number in the upper left: $\sqrt[n]{a}$, read as "the nth root of a." The relationship between radical notation and exponent notation can be described by this equation: $\sqrt[n]{a} = a^{\frac{1}{n}}$. The two special cases of $n = 2$ and $n = 3$ are called square roots and cube roots. If there is no number to the upper left, it is understood to be a square root ($n = 2$). Nearly all of the roots you encounter will be square roots. A square root is the same as a number raised to the one-half power. When we say that a is the square root of b ($a = \sqrt{b}$), we mean that a multiplied by itself equals b: ($a \times a = b$).

A perfect square is a number that has an integer for its square root. There are 10 perfect squares from 1 to 100: 1, 4, 9, 16, 25, 36, 49, 64, 81, 100 (the squares of integers 1 through 10).

Scientific notation is a way of writing large numbers in a shorter form. The form $a \times 10^n$ is used in scientific notation, where a is greater than or equal to 1, but less than 10, and n is the number of places the decimal must move to get from the original number to a. Example: The number 230,400,000 is cumbersome to write. To write the value in scientific notation, place a decimal point between the first and second numbers, and include all digits through the last non-zero digit ($a = 2.304$). To find the appropriate power of 10, count the number of places the decimal point had to move ($n = 8$). The number is positive if the decimal moved to the left, and negative if it moved to the right. We can then write 230,400,000 as 2.304×10^8. If we look instead at the number 0.00002304, we have the same value for a, but this time the decimal moved 5 places to the right ($n = -5$). Thus, 0.00002304 can be written as 2.304×10^{-5}. Using this notation makes it simple to compare very large or very small numbers. By comparing exponents, it is easy to see that 3.28×10^4 is smaller than 1.51×10^5, because 4 is less than 5.

Factors and multiples

Factors are numbers that are multiplied together to obtain a product. For example, in the equation $2 \times 3 = 6$, the numbers 2 and 3 are factors. A prime number has only two factors (1 and itself), but other numbers can have many factors.
A common factor is a number that divides exactly into two or more other numbers. For example, the factors of 12 are 1, 2, 3, 4, 6, and 12, while the factors of 15 are 1, 3, 5, and 15. The common factors of 12 and 15 are 1 and 3.
A prime factor is also a prime number. Therefore, the prime factors of 12 are 1, 2, and 3. For 15, the prime factors are 1, 3, and 5.

The greatest common factor (GCF) is the largest number that is a factor of two or more numbers. For example, the factors of 15 are 1, 3, 5, and 15; the factors of 35 are 1, 5, 7, and 35. Therefore, the greatest common factor of 15 and 35 is 5.
The least common multiple (LCM) is the smallest number that is a multiple of two or more numbers. For example, the multiples of 3 include 3, 6, 9, 12, 15, etc.; the multiples of 5 include 5, 10, 15, 20, etc. Therefore, the least common multiple of 3 and 5 is 15.

Solving systems of equations

Systems of Equations are a set of simultaneous equations that all use the same variables. A solution to a system of equations must be true for each equation in the system. *Consistent Systems* are those with at least one solution. *Inconsistent Systems* are systems of equations that have no solution.

To solve a system of linear equations by *substitution*, start with the easier equation and solve for one of the variables. Express this variable in terms of the other variable. Substitute this expression in the other equation, and solve for the other variable. The solution should be expressed in the form (x, y). Substitute the values into both of the original equations to check your answer. Consider the following problem.

Solve the system using substitution:

$$x + 6y = 15$$
$$3x - 12y = 18$$

Solving the first equation for x:

$$x = 15 - 6y$$

Substitute this value in place of x in the second equation, and solve for y:

$$3(15 - 6y) - 12y = 18$$
$$45 - 18y - 12y = 18$$
$$30y = 27$$
$$y = \frac{27}{30} = \frac{9}{10} = 0.9$$

Plug this value for y back into the first equation to solve for x:

$$x = 15 - 6(0.9) = 15 - 5.4 = 9.6$$

Check both equations if you have time:

$$9.6 + 6(0.9) = 9.6 + 5.4 = 15$$
$$3(9.6) - 12(0.9) = 28.8 - 10.8 = 18$$

Therefore, the solution is $(9.6, 0.9)$.

To solve a system of equations using *elimination*, begin by rewriting both equations in standard form $Ax + By = C$. Check to see if the coefficients of one pair of like variables add to zero. If not, multiply one or both of the equations by a non-zero number to make one set of like variables add to zero. Add the two equations to solve for one of the variables. Substitute this value into one of the original equations to solve for the other variable. Check your work by substituting into the other equation. Next we will solve the same problem as above, but using the addition method.

Solve the system using elimination:

$$x + 6y = 15$$
$$3x - 12y = 18$$

If we multiply the first equation by 2, we can eliminate the y terms:

$$2x + 12y = 30$$
$$3x - 12y = 18$$

Add the equations together and solve for x:

$$5x = 48$$
$$x = \frac{48}{5} = 9.6$$

Plug the value for x back in to either of the original equations and solve for y:

$$9.6 + 6y = 15$$
$$y = \frac{15 - 9.6}{6} = 0.9$$

Check both equations if you have time:

$$9.6 + 6(0.9) = 9.6 + 5.4 = 15$$
$$3(9.6) - 12(0.9) = 28.8 - 10.8 = 18$$

Therefore, the solution is (9.6, 0.9).

Angles

An angle is formed when two lines or line segments meet at a common point. It may be a common starting point for a pair of segments or rays, or it may be the intersection of lines. Angles are represented by the symbol ∠.

The vertex is the point at which two segments or rays meet to form an angle. If the angle is formed by intersecting rays, lines, and/or line segments, the vertex is the point at which four angles are formed. The pairs of angles opposite one another are called vertical angles, and their measures are equal. In the figure below, angles ABC and DBE are congruent, as are angles ABD and CBE.

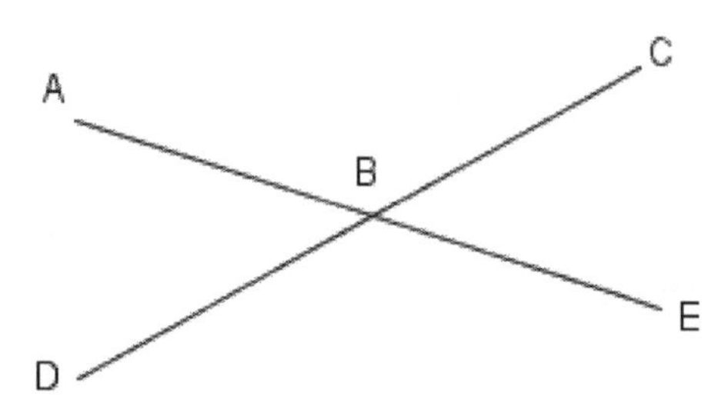

An acute angle is an angle with a degree measure less than 90°.
A right angle is an angle with a degree measure of exactly 90°.
An obtuse angle is an angle with a degree measure greater than 90° but less than 180°.
A straight angle is an angle with a degree measure of exactly 180°. This is also a semicircle.
A reflex angle is an angle with a degree measure greater than 180° but less than 360°.
A full angle is an angle with a degree measure of exactly 360°.

Two angles whose sum is exactly 90° are said to be complementary. The two angles may or may not be adjacent. In a right triangle, the two acute angles are complementary.

Two angles whose sum is exactly 180° are said to be supplementary. The two angles may or may not be adjacent. Two intersecting lines always form two pairs of supplementary angles. Adjacent supplementary angles will always form a straight line.

Circles

The center is the single point inside the circle that is equidistant from every point on the circle. (Point *O* in the diagram below.)

The radius is a line segment that joins the center of the circle and any one point on the circle. All radii of a circle are equal. (Segments *OX*, *OY*, and *OZ* in the diagram below.)

The diameter is a line segment that passes through the center of the circle and has both endpoints on the circle. The length of the diameter is exactly twice the length of the radius. (Segment *XZ* in the diagram below.)

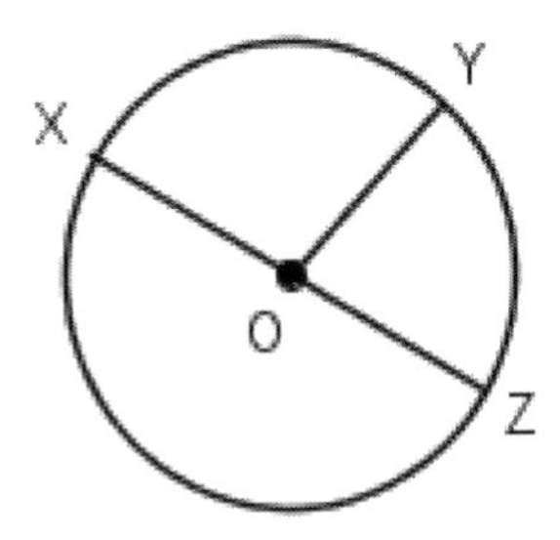

Triangles

A triangle is a polygon with three sides and three angles. Triangles can be classified according to the length of their sides or magnitude of their angles.

An acute triangle is a triangle whose three angles are all less than 90°. If two of the angles are equal, the acute triangle is also an isosceles triangle. If the three angles are all equal, the acute triangle is also an equilateral triangle.

A right triangle is a triangle with exactly one angle equal to 90°. All right triangles follow the Pythagorean Theorem. A right triangle can never be acute or obtuse.

An obtuse triangle is a triangle with exactly one angle greater than 90°. The other two angles may or may not be equal. If the two remaining angles are equal, the obtuse triangle is also an isosceles triangle.

An equilateral triangle is a triangle with three congruent sides. An equilateral triangle will also have three congruent angles, each 60°. All equilateral triangles are also acute triangles.

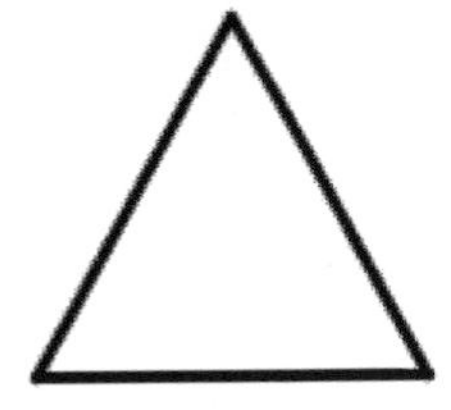

An isosceles triangle is a triangle with two congruent sides. An isosceles triangle will also have two congruent angles opposite the two congruent sides.

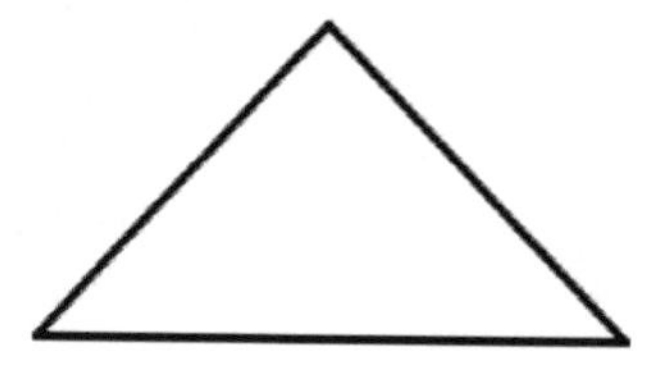

A scalene triangle is a triangle with no congruent sides. A scalene triangle will also have three angles of different measures. The angle with the largest measure is opposite the longest side, and the angle with the smallest measure is opposite the shortest side.

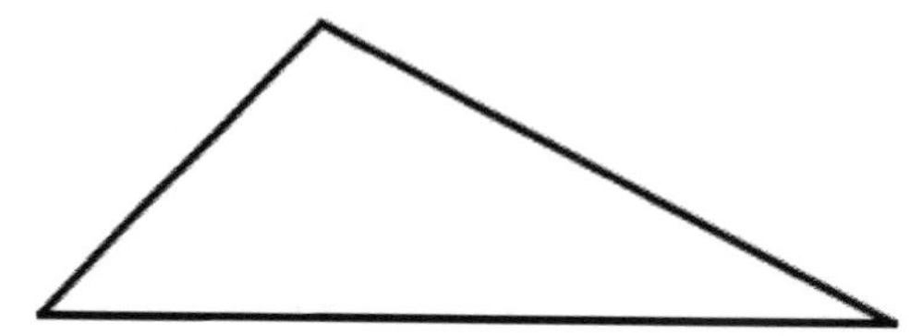

The Triangle Inequality Theorem states that the sum of the measures of any two sides of a triangle is always greater than the measure of the third side. If the sum of the measures of two sides were equal to the third side, a triangle would be impossible because the two sides would lie flat across the third side and there would be no vertex. If the sum of the measures of two of the sides was less than the third side, a closed figure would be impossible because the two shortest sides would never meet.

Similar triangles are triangles whose corresponding angles are congruent to one another. Their corresponding sides may or may not be equal, but they are proportional to one another. Since the angles in a triangle always sum to 180°, it is only necessary to determine that two pairs of corresponding angles are congruent, since the third will be also in that case.

Congruent triangles are similar triangles whose corresponding sides are all equal. Congruent triangles can be made to fit on top of one another by rotation, reflection, and/or translation. When trying to determine whether two triangles are congruent, there are several criteria that can be used.

Practice Questions

1. If the volume of a cube is 8 cm^3, what is the length of the cube?
   a. 1 cm
   b. 2 cm
   c. 3 cm
   d. 4 cm
   e. 8 cm

2. Simplify the following expression:
$(2x^2 + 3) (2x - 1)$
   a. $4x^3 - 2x^2 + 6x - 3$
   b. $2x^2 + 6x - 3$
   c. $4x^3 - 2x^2 + 6x + 3$
   d. $4x^3 - 2x^2 - 6x - 3$
   e. $2x^2 - 6x + 3$

3. Simplify the following expression
$(2x^4y^7m^2z) * (5x^2y^3m^8)$
   a. $10x^6y^9m^{10}z$
   b. $7x^6y^{10}m^{10}z$
   c. $10x^5y^{10}m^{10}z$
   d. $10x^6y^{10}m^{10}z$
   e. $7x^5y^9m^{10}z$

4. A classroom contains 13 boys and 18 girls. If a student's name is chosen randomly, what is the probability it will be a girl's name?
   a. 36%
   b. 42%
   c. 58%
   d. 72%
   e. 84%

5. If $x - 9 = 2x + 10$, what is the value of x?
   a. -19
   b. 19
   c. 6.3
   d. -6.3
   e. none of the above

Practice Answers

1. B: The volume of a cube is calculated by cubing the length, width, or height of the cube (the value for all three of these is the same.

Therefore, the volume of a cube equals = $length^3$
In this case $8cm^3 = x * x * x$, where x can represent the length of the cube.

To find the length, we must figure out which number cubed equals 8.
The answer is 2cm: $2cm * 2cm * 2cm = 8cm^3$

2. A: Use the FOIL (first, outside, inside, last) to expand the expression:
$4x^3 - 2x^2 + 6x - 3$
There are no like terms, so the expression cannot be simplified any further.

3. D: To simplify this expression, the law of exponents that states that $x^m * x^n = x^{m+n}$ must be observed.
$10x^{4+2}y^{7+3}m^{2+8}z$

Therefore, $10x^6y^{10}m^{10}z$ is the simplified expression.

4. C: First, find the total number of students in the classroom: $13 + 18 = 31$
There is an 18 in 31 chance that a name chosen randomly will be a girl's name.

To express this as a percentage, divide 18 by 31, then multiply that number by 100:
$18/31 * 100\% = 58\%$

5. A: First, gather all of the terms that contain an x on the left side of the equation to make it easier to solve:
$x - 2x - 9 = 10$
$-x - 9 = 10$

Then, add nine to both sides to isolate the x:
$-x - 9 + 9 = 10 + 9$
$-x = 19$

Finally, divide by -1 to solve for x:
$-x/-1 = 19/-1$
$x = -19$

Reading Comprehension

What do reading comprehension questions look like?

The questions in this section will follow the typical format of reading comprehension questions on standardized tests. You'll be given a passage to read, several paragraphs in length, and then be shown several questions about the passage. Each question will have five possible answers to choose from; only one answer will be correct.

What are they testing?

The questions in this section are testing your ability to read and understand written material. Most people taking the AFOQT will have a level of reading ability that's quite a bit higher than the reading ability of the average American, and the passages in this section will reflect that fact. They will tend to be written on the level of articles in academic and scientific publications, and not at the level of articles one reads in most magazines or newspapers, or on most websites.

However, don't take this to mean that you'll need any specialized scientific or technical knowledge to do well on these questions; you won't. Each passage, while written on a higher level than average, and possibly on a scientific or technical subject, will assume no specialized knowledge on the part of the reader about the subject matter.

Some questions will test your ability to remember or quickly locate facts in the passage, although the questions will generally not use the same words or phrases used in the passage. Other questions will require you to make judgments about what you've read, such as choosing a statement the author would agree or disagree with, or deciding what the author's main point was, or what the author's purpose in writing the passage was.

How can I prepare?

The vast majority of Americans engage in very little reading these days, beyond texts and social media. Needless to say, that kind of reading won't suffice to prepare you for the AFOQT. Even if you actually do some legitimate reading from time to time, odds are that you read much less than the average aspiring Air Force officer of 30 years ago did, and you too definitely need to brush up on your reading comprehension skills.

In either case, you'll want to set aside regular practice sessions where you read higher level reading passages than you're used to, and then ask yourself a lot of questions after each passage.

- What is the author's main point?
- What kind of statements would he be likely to agree with or disagree with?
- Is the passage educational, persuasive, argumentative, etc.?
- What are some of the secondary points the author makes?
- Is the article fact based, or merely stating an opinion?
- Has the author made a good case?
- If you think he has, what do you base this on?
- If not, where is his argument or article weak?

Libraries have a lot of great resources you can take advantage of to help you improve your reading comprehension skills.

Practice Questions

Sample Passage

Historically, the term pilot error has been used to describe an accident in which an action or decision made by the pilot was the cause or a contributing factor that led to the accident. This definition also includes the pilot's failure to make a correct decision or take proper action. From a broader perspective, the phrase human factors related more aptly describes these accidents. A single decision or event does not lead to an accident, but a series of events and the resultant decisions together form a chain of events leading to an outcome.

In his article *Accident-Prone Pilots*, Dr. Patrick R. Veillette uses the history of Captain Everyman to demonstrate how aircraft accidents are caused more by a chain of poor choices rather than one single poor choice. In the case of Captain Everyman, after a gear-up landing accident, he became involved in another accident while taxiing a Beech 58P Baron out of the ramp. Interrupted by a radio call from the 17-11 dispatcher, Everyman neglected to complete the fuel crossfeed check before taking off. Everyman, who was flying solo, left the right-fuel selector in the cross-feed position. Once aloft and cruising, he noticed a right roll tendency and corrected with aileron trim. He did not realize that both engines were feeding off the left wing's tank, making the wing lighter.

After two hours of flight, the right engine quit when Everyman was flying along a deep canyon gorge. While he was trying to troubleshoot the cause of the right engine's failure, the left engine quit. Everyman landed the aircraft on a river sand bar but it sank into ten feet of water.

Several years later, Everyman flew a de Havilland Twin Otter to deliver supplies to a remote location. When he returned to home base and landed, the aircraft veered sharply to the left, departed the runway, and ran into a marsh 375 feet from the runway. The airframe and engines sustained considerable damage. Upon inspecting the wreck, accident investigators found the nose wheel steering tiller in the fully deflected position. Both the after-takeoff and before-landing checklists required the tiller to be placed in the neutral position. Everyman had overlooked this item.

Now, is Everyman accident prone or just unlucky? Skipping details on a checklist appears to be a common theme in the preceding accidents. While most pilots have made similar mistakes, these errors were probably caught prior to a mishap due to extra margin, good warning systems, a sharp copilot, or just good luck. What makes a pilot less prone to accidents?

The successful pilot possesses the ability to concentrate, manage workloads, monitor and perform several simultaneous tasks. Some of the latest psychological screenings used in aviation test applicants for their ability to multitask, measuring both accuracy, as well as the individual's ability to focus attention on several subjects simultaneously. The FAA oversaw an extensive research study on the similarities and dissimilarities of accident-free pilots and those who were not. The project surveyed over 4,000 pilots, half of whom had clean records while the other half had been involved in an accident.

Five traits were discovered in pilots prone to having accidents. These pilots:

- Have disdain toward rules.
- Have very high correlation between accidents on their flying records and safety violations on their driving records.
- Frequently fall into the thrill and adventure seeking personality category.
- Are impulsive rather than methodical and disciplined, both in their information gathering and in the speed and selection of actions to be taken.
- Show a disregard for or under-utilization of outside sources of information, including copilots, flight attendants, flight service personnel, flight instructors, and air traffic controllers.

(Questions on the following page)

1. The primary purpose of the passage is to
   a. criticize Captain Everyman.
   b. advocate for more comprehensive pilot training.
   c. reduce flight accidents.
   d. entertain the reader with a story of pilot incompetence.
   e. describe some characteristics that correlate with flight accidents.

2. Which of the following statements about Captain Everyman is NOT supported by the passage?
   a. Captain Everyman is a reckless thrill-seeker.
   b. Captain Everyman is easily distracted.
   c. Captain Everyman is not methodical in his work.
   d. Captain Everyman can fly a variety of aircraft.
   e. Captain Everyman does not double-check his work.

3. Why does the author prefer to describe the main cause of accidents as human factors rather than pilot error?
   a. Because most accidents are caused by multiple poor decisions rather than a single error.
   b. Because most accidents are caused by faulty equipment.
   c. Because most accidents can be traced to a single mistake.
   d. Because many accidents are caused by miscommunication with the control tower.
   e. Because pilots can be subject to criminal charges of negligence.

4. In the fifth paragraph, *prone* most nearly means
   a. immobile
   b. supine
   c. likely to have
   d. lying down
   e. encouraging

5. With which one of the following claims about pilots would the author most likely agree?
   a. Pilots should not be allowed to fly solo until they are thirty years old.
   b. A good pilot and a good short-order cook have many of the same skills.
   c. Pilots should be restricted to flying one type of plane.
   d. Pilots should never fly solo.
   e. Some pilots never make mistakes.

Practice Answers

1. E. Describing some characteristics that correlate with flight accidents is the primary purpose of the passage. The author explains how most flight accidents are the result of several mistakes in sequence, and how inattention, inability to multi-task, and carelessness are common characteristics of the accident-prone pilot.

2. A. The passage presents a critical portrait of Captain Everyman, but it never directly suggests that he is a reckless thrill-seeker. The passage does state that this personality type is correlated with flight accidents, but, in the example scenarios, the causes of Captain Everyman's accidents are carelessness and distraction rather than recklessness.

3. A. The author prefers saying that the main cause of accidents is human factors, and not simply pilot error, because accidents are rarely the result of a single mistake, but rather are typically caused by a series of mistakes, oversights, or general carelessness over a period of time. The author makes this point in the first paragraph of the passage.

4. C. In the fifth paragraph, *prone* most nearly means *likely to have*. The author is discussing pilots who are accident prone, meaning that they are more likely to have accidents. Prone can also mean *lying down,* or *supine*, but it does not have that meaning in this context.

5. B. The author would most likely agree that a good pilot and a good short-order cook have many of the same skills. Specifically, the author emphasizes the importance of multi-tasking as a pilot, a skill which would also be important for a short-order cook, who must prepare several different meals at the same time. There is no indication in the passage that age or flying different types of aircraft is correlated with accidents, and there is no argument that pilots should never fly solo. Finally, the author explicitly states that all pilots make mistakes, though some are better than others at correcting them.

Situational Judgment

What do situational judgment questions look like?

All situational judgment questions will have a similar format. Test takers will be given a scenario which requires some action to be taken to solve a problem that has come up, usually involving interpersonal and/or official relationships between an officer and his subordinates and/or his superiors. The test taker will then be shown several possible actions that could be taken, and will be told to select both the **most effective** and the **least effective** of the actions listed.

What are situational judgment questions testing?

This is a newer section of the AFOQT, added to make the test a better predictor of success as an Air Force officer. These questions are primarily focused on testing a person in the areas of judgment and self-sufficient decision making abilities. In order to do well on this section, test takers will need to show that they can lead subordinates and solve problems independently by using their core competencies of resource management, communication, innovation, mentoring, leadership, professionalism, and integrity.

How can I prepare?

Situational judgment questions aren't the kind of questions that lend themselves easily to preparation, as there really isn't any material for a person to review and memorize. However, you should keep in mind the qualities listed above (resource management, communication, innovation, mentoring, leadership, professionalism, and integrity) when answering questions. Your answers should reflect these qualities as much as possible. Also, you should avoid choosing any answer which involves going to a superior for advice or help unless there are no other viable options, or discussing a person's shortcomings behind their back no matter their rank. Officers are expected to be resourceful men and women of character.

Physical Science

What do physical science questions look like?

Physical science questions primarily test your understanding of scientific terms. You won't be asked to perform complex physics calculations or balance a chemical reaction. The purpose of this section is to make sure you paid attention in high school science and retained some of the general concepts.

How can I prepare?

The best way to prepare for these questions is to brush up on your science terminology and concepts. We've included a glossary of terms here to give you head start, but to be more thorough, a good idea would be to find a high school physical science textbook and look through the full glossary in there.

A

Absolute zero: The lowest possible temperature (-273.15°C).
Atmospheric pressure: The pressure exerted by the gases in the air. Units of measurement are kilopascals (kPa), atmospheres (atm), millimeters of mercury (mm Hg) and Torr. Standard atmospheric pressure is 100 kPa, 1atm, 760 mm Hg or 760 Torr.
Atom: The smallest particle of an element; a nucleus and its surrounding electrons.
Atomic mass: The mass of an atom measured in atomic mass units (amu). An atomic mass unit is equal to one-twelfth of the atom of carbon-12. Atomic mass is now more generally used instead of atomic weight. Example: the atomic mass of chlorine is about 35 amu.
Atomic number: Also known as proton number, it is the number of electrons or the number of protons in an atom. Example: the atomic number of gold is 79.
Atomic weight: A common term used to mean the average molar mass of an element. This is the mass per mole of atoms. Example: the atomic weight of chlorine is about 35 g/mol.

B

Boiling point: The temperature at which a substance undergoes a phase change from a liquid to a gas.

C

Celsius scale (°C): A temperature scale on which the freezing point of water is at 0 degrees and the normal boiling point at standard atmospheric pressure is 100 degrees.
Change of state: A change between two of the three states of matter, solid, liquid and gas. Example: when water evaporates it changes from a liquid to a gaseous state.
Compound: A chemical consisting of two or more elements chemically bonded together. Example: Calcium can combine with carbon and oxygen to make calcium carbonate ($CaCO_3$), a compound of all three elements.
Condensation: The formation of a liquid from a gas. This is a change of state, also called a phase change.
Conduction: (i) the exchange of heat (heat conduction) by contact with another object, or (ii) allowing the flow of electrons (electrical conduction).
Convection: The exchange of heat energy with the surroundings produced by the flow of a fluid due to being heated or cooled.

D

Decay (radioactive decay): The way that a radioactive element changes into another element due to loss of mass through radiation. Example: uranium 238 decays with the loss of an alpha particle to form thorium 234.
Density: The mass per unit volume (e.g. g/cm^3).
Diffusion: The slow mixing of one substance with another until the two substances are evenly mixed. Mixing occurs because of differences in concentration within the mixture. Diffusion works rapidly with gases, very slowly with liquids.
Dissolve: To break down a substance in a solution without causing a reaction.

E

Electrical potential: The energy produced by an electrochemical cell and measured by the voltage or electromotive force (emf).
Electron: A tiny, negatively charged particle that is part of an atom. The flow of electrons through a solid material such as a wire produces an electric current.
Element: A substance that cannot be decomposed into simpler substance by chemical means. Examples: calcium, iron, gold.
Explosive: A substance which, when a shock is applied to it, decomposes very rapidly, releasing a very large amount of heat and creating a large volume of gases as a shock wave.

F

Fluid: Able to flow; either a liquid or a gas.
Freezing point: The temperature at which a substance undergoes a phase change from a liquid to a solid. It is the same temperature as the melting point.

G

Gamma rays: Waves of radiation produced as the nucleus of a radioactive element rearranges itself into a tighter cluster of protons and neutrons. Gamma rays carry enough energy to damage living cells.
Gas/gaseous phase: A form of matter in which the molecules form no definite shape and are free to move about to uniformly fill any vessel they are put in. A gas can easily be compressed into a much smaller volume.
Group: A vertical column in the Periodic Table. There are eight groups in the table. Their numbers correspond to the number of electrons in the outer shell of the atoms in the group. Example: Group 2 contains beryllium, magnesium, calcium, strontium, barium and radium.

H

Half-life: The time it takes for the radiation coming from a sample of a radioactive element to decrease by half.
Heat: The energy that is transferred when a substance is at a different temperature to that of its surroundings.
Heat capacity: The ratio of the heat supplied to a substance, compared with the rise in temperature that is produced.
Heat of combustion: The amount of heat given off by a mole of a substance during combustion. This heat is a property of the substance and is the same no matter what kind of combustion is involved. Example: heat of combustion of carbon is 94.05 kcal (x 4.18 kJ/kcal = 393.1 kJ).

I

Ion: An atom, or group of atoms, that has gained or lost one or more electrons and so developed an electrical charge. Ions behave differently from electrically neutral atoms and molecules. They can move in an electric field, and they can also bind strongly to solvent molecules such as water. Positively charged ions are called cations; negatively charged ions are called anions. Ions can carry an electrical current through solutions.
Isotope: One of two or more atoms of the same element that have the same number of protons in their nucleus (atomic number), but which have a different number of neutrons (atomic mass). Example: carbon-12 and carbon-14.

K

Kinetic energy: The energy an object has by virtue of its being in motion.

L

Latent heat: The amount of heat that is absorbed or released during the process of changing state between gas, liquid or solid. For example, heat is absorbed when a substance melts and it is released again when the substance solidifies.
Liquid/liquid phase: A form of matter that has a fixed volume but no fixed shape.

M

Mass: The amount of matter in an object. In everyday use the word weight is often used (somewhat incorrectly) to mean mass.
Matter: Anything that has mass and takes up space.
Melting point: The temperature at which a substance changes state from a solid phase to a liquid phase. It is the same as freezing point.
Metal: A class of elements that is a good conductor of electricity and heat, has a metallic luster, is malleable and ductile, forms cations and has oxides that are bases. Metals are formed as cations held together by a sea of electrons. A metal may also be an alloy of these elements. Example: sodium, calcium, gold.
Mixture: A material that can be separated into two or more substances using physical means. Example: a mixture of copper (II) sulfate and cadmium sulfide can be separated by filtration.
Mole: 1 mole is the amount of a substance which contains Avogadro's number (about 6×10^{23}) of particles. Example: 1 mole of carbon-12 weighs exactly 12 g.
Molecule: A group of two or more atoms held together by chemical bonds. Example: O_2.

N

Neutron: A particle inside the nucleus of an atom that is neutral and has no charge.
Newton (N): The unit of force required to give one kilogram an acceleration of one meter per second every second (1 m/s^2).
Noble gases: The members of Group 8 of the Periodic Table: helium, neon, argon, krypton, xenon, and radon. These gases are almost entirely unreactive.
Nucleus: The small, positively charged particle at the centre of an atom. The nucleus is responsible for most of the mass of an atom.

P

Period: A row in the Periodic Table.
Periodic Table: A chart organizing elements by atomic number and chemical properties into groups and periods.
Phase: A particular state of matter. A substance may exist as a solid, liquid or gas and may change between these phases with addition or removal of energy. Examples: ice, liquid and vapor are the three phases of water. Ice undergoes a phase change to water when heat energy is added.

Photon: A parcel of light energy.
Potential energy: The energy an object has by virtue of its position or orientation, most commonly its height above some reference point, or amount of compression as with a spring.
Pressure: The force per unit area measured in Pascals.
Proton: A positively charged particle in the nucleus of an atom that balances out the charge of the surrounding electrons.

R

Radiation: The exchange of energy with the surroundings through the transmission of waves or particles of energy. Radiation is a form of energy transfer that can happen through space; no intervening medium is required (as would be the case for conduction and convection).

S

Solid/solid phase: A rigid form of matter which maintains its shape, whatever its container.

Table Reading

What are these questions testing?

The table reading section tests your ability to quickly and accurately locate information stored in a table. The questions require you to find a particular number in a table given a set of coordinates.

What is the question format?

The questions will be given in groups of around 5 questions, where each group will refer to a table of numbers, with column and row headers. Each question will give an ordered pair of numbers that indicate the location in the table where the correct answer can be found. The ordered pair is given in the form (x, y), where x is the column number and y is the row number.

The questions will be presented like this:

1. (3, -2)

a. 56
b. 39
c. 64
d. 11
e. 92

	-3	-2	-1	0	1	2	3
3	41	39	84	77	35	42	37
2	75	57	95	16	93	16	15
1	34	54	50	89	26	19	94
0	66	89	65	23	13	42	20
-1	15	97	86	76	76	58	92
-2	80	92	78	52	90	11	56
-3	88	81	61	79	35	64	52

To answer the question, look at the ordered pair. It indicates that the number you are looking for is in the column labeled 3, and the row labeled -2. In this table, that number is 56, answer choice A.

What strategies can I use to answer the questions quickly and accurately?

The best way to approach these questions is methodically. Take the first number in the ordered pair and find it on the column headers. Keep your finger on that number while you take the second number of the ordered pair and locate it on the row headers. Put another finger on that number. Drag the first finger straight down the column until you get to the row that your other finger is on. The number at the intersection of the indicated column and row is your answer.

If you find yourself staring at the table for too long trying to be sure that you've selected the right number, you may find it helpful to draw lines on the graph, or take two pieces of scratch paper and line up their edges with the column and row numbers given in the question so that the number you are looking for appears at the corner where the two pieces of paper come together.

What are the common mistakes to avoid?

Since the process for answering these questions is very straightforward, most errors are the result of trying to go too quickly. The more you practice these sorts of questions, the faster you will be able to accurately answer them. Once you've practiced for a while, you'll be able to get a feel for how fast you can go while accurately answering all the questions. On test day, force yourself to go no faster than this.

One common mistake made on these questions is taking the ordered pair in the question to be ordinal coordinates rather than numbers referencing the column and row labels. For example, suppose a question asks for the ordered pair (1, 2). Under the pressure of the test, many people will instinctively go to the first column and second row, and take that number. Don't succumb to pressure on the test. Just follow the procedure and work through the questions at your ideal speed.

Practice Questions

For each question, select the number that appears in the table at the given coordinates. Recall that the first number in the ordered pair gives the column number, and the second gives the row number. For instance, the ordered pair (0, -2) refers to the number in column 0, row -2, which is 84 in the table below.

Use the table on the right to answer questions 1-5.

	-3	-2	-1	0	1	2	3
3	89	57	70	68	11	95	40
2	85	28	75	82	63	42	58
1	85	20	16	52	62	87	87
0	25	83	78	21	73	11	31
-1	30	72	73	51	91	30	70
-2	16	82	32	84	28	91	63
-3	74	19	96	38	49	17	25

1. (-1, 1)
a. 63
b. 84
c. 16
d. 51
e. 87

2. (-2, 3)
a. 57
b. 21
c. 25
d. 58
e. 51

3. (3, -3)
a. 58
b. 25
c. 89
d. 16
e. 21

4. (-2, 2)
a. 28
b. 96
c. 11
d. 72
e. 49

5. (-1, -2)
a. 17
b. 95
c. 32
d. 42
e. 25

Use the table on the right to answer questions 6-10.

	-3	-2	-1	0	1	2	3
3	94	49	12	84	91	92	33
2	23	79	99	97	33	51	22
1	50	92	16	12	74	86	53
0	59	11	55	72	86	29	65
-1	14	66	14	34	16	97	17
-2	27	37	82	52	18	39	43
-3	79	39	96	22	87	98	54

6. (1, 1)
a. 28
b. 16
c. 42
d. 73
e. 74

7. (-2, 1)
a. 49
b. 92
c. 28
d. 91
e. 57

8. (0, 2)
a. 30
b. 25
c. 20
d. 91
e. 97

9. (3, -3)
a. 54
b. 82
c. 87
d. 95
e. 20

10. (-2, -1)
a. 82
b. 63
c. 52
d. 72
e. 66

Practice Answers

1. C

2. A

3. B

4. A

5. C

6. E

7. B

8. E

9. A

10. E

Instrument Comprehension

What are these questions testing?

These questions are designed to test your familiarity with and understanding of common instruments.

What is the question format?

The question format will vary depending on the types of instruments that the test covers. However, this section of the book will focus on the compass and the artificial horizon, two instruments commonly used in airplanes.

Test questions related to these instruments will illustrate a compass and an artificial horizon inside an airplane cockpit. Based on the readings of both of these instruments, you will have to determine the position and orientation of the airplane.

How do I read the compass and artificial horizon instruments?

The compass, a relatively intuitive instrument with which many people are familiar, shows which direction a person or vehicle is facing. When a person is facing north, for example, the needle on the compass points toward the "N." If the person is facing a direction between south and southeast, the needle will point between "S" and "SE."

The artificial horizon is an instrument that shows how the nose and wings of a plane are tilted. For most people, the artificial horizon is less intuitive and less familiar than the compass. However, if you imagine yourself actually flying in a plane, the artificial horizon becomes easier to read and understand.

The artificial horizon has two components that illustrate how the nose of an airplane is tilted with respect to the ground: the miniature wings and the horizon bar. The miniature wings represent the actual wings of the aircraft, and the horizon bar represents the horizon, the imaginary line that divides the ground and the sky from the pilot's point of view. When the miniature wings are level with the horizon bar, the plane is level. When the miniature wings are above the horizon bar, the plane is tilted upward, and when the miniature wings are below the horizon bar, the plane is tilted downward. These categories of nose tilt are shown in the drawing below.

To illustrate how the wings are tilted from side to side, the artificial horizon instrument also has a dial with degree marks representing the bank angle. A needle on the dial indicates the exact bank angle, and the horizon bar is tilted accordingly, as shown in the picture below. If the left wing of the plane is tilted downward, the needle will be to the right of the center of the dial; if the right wing is tilted downward; the needle will be to the left of the center. Note that the tilted horizon bar reflects the pilot's point of view: if the left wing of the plane is tilted downward, the horizon will appear to be tilted in the opposite direction.

To answer the questions, you will have to use information from both the compass and artificial horizon to determine how the plane is oriented. If the plane is flying north, it will appear to fly into the page in the illustrations.

For example, based on the compass and artificial horizon shown below, which of the answer choices represents the orientation of the plane?

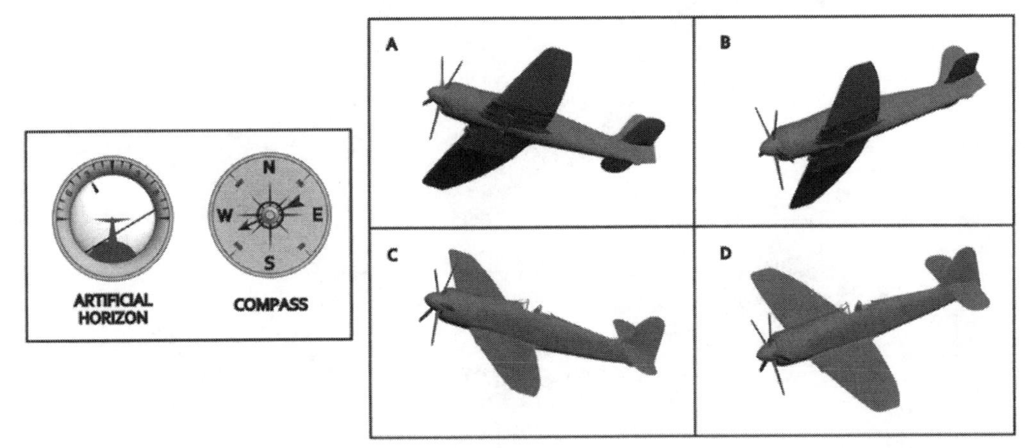

ANSWER: The answer is A. First, notice that the compass is pointing in a west-southwest direction. If a north-flying plane is facing into the page, then a westbound plane will be facing left. The compass indicates that the plane is flying somewhere between west and southwest, so the illustration will show a plane that appears to be facing left and just slightly out of the page.

Second, notice that the miniature wings in the artificial horizon are above the horizon line, and the needle on the dial is to the left of the center. From this information, you know that 1) the nose of the plane is tilted upward, and 2) the left wing of the plane is tilted upward, and the right wing is tilted downward. Because only the plane illustrated in choice A fits this description, it is the correct answer.

How can I improve my ability to read the compass and artificial horizon?

Most people don't encounter these instruments on an everyday basis, so the best way to improve your ability to read them is simply to do the practice question in this section. However, although the artificial horizon is not commonly found outside of aircraft, you might want to practice using a real compass if you are still having trouble with these questions. You can find a reasonably-priced compass at most outdoor or sporting goods stores.

Practice Questions

1. Which of the answer choices represents the orientation of the plane?

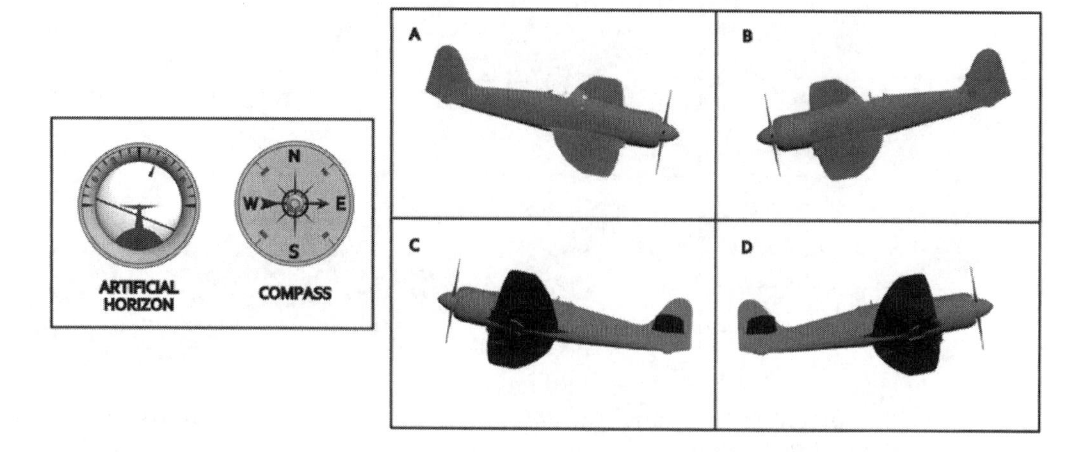

2. Which of the answer choices represents the orientation of the plane?

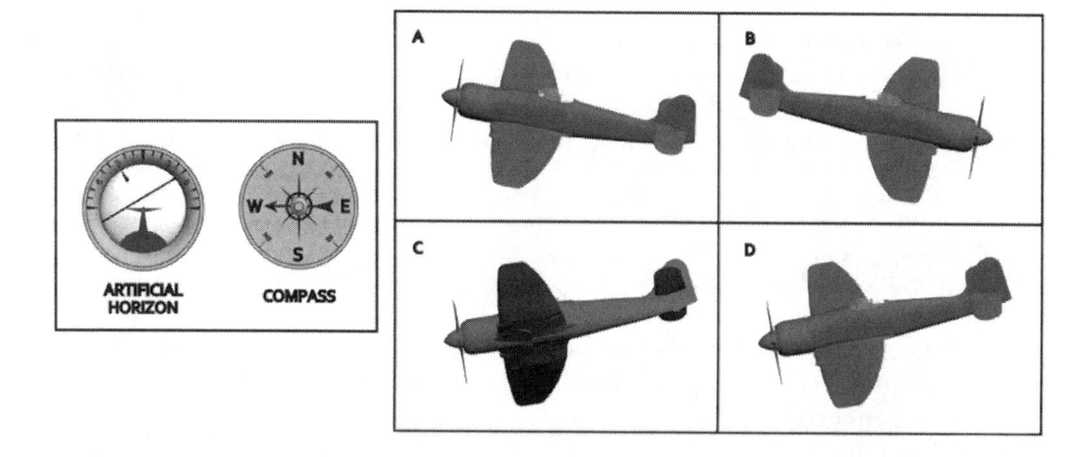

3. Which of the answer choices represents the orientation of the plane?

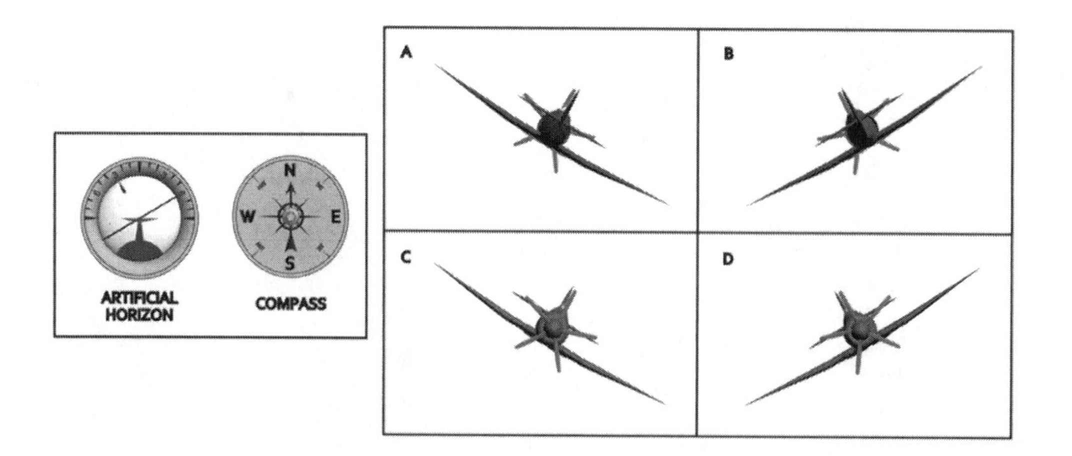

4. Which of the answer choices represents the orientation of the plane?

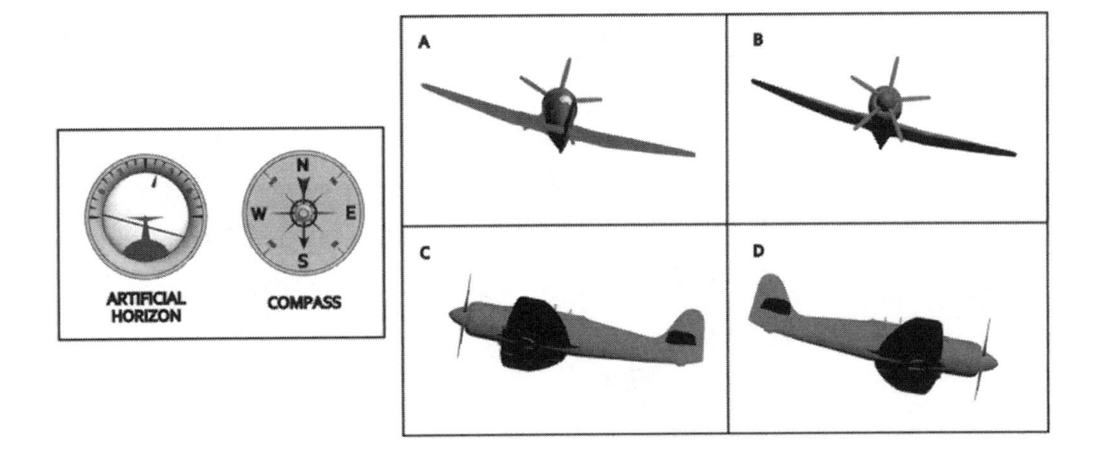

5. Which of the answer choices represents the orientation of the plane?

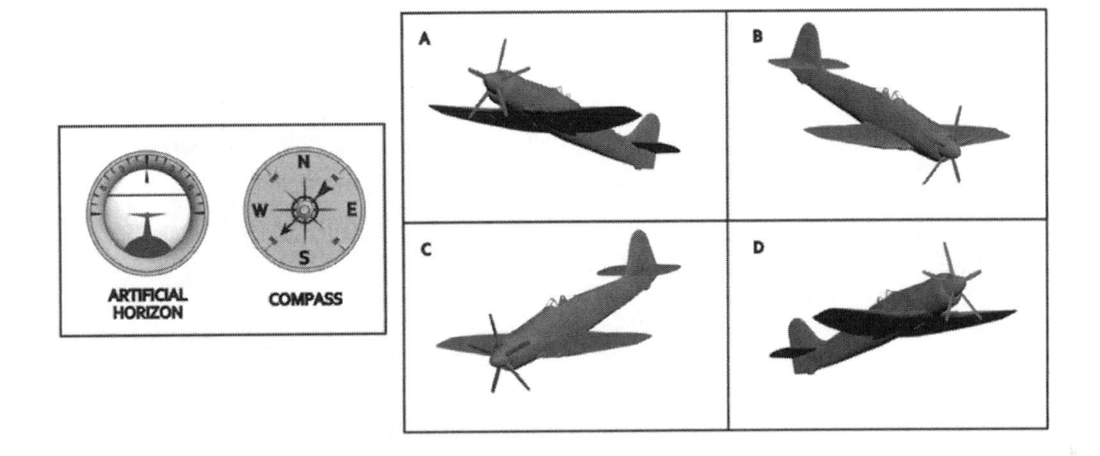

6. Which of the answer choices represents the orientation of the plane?

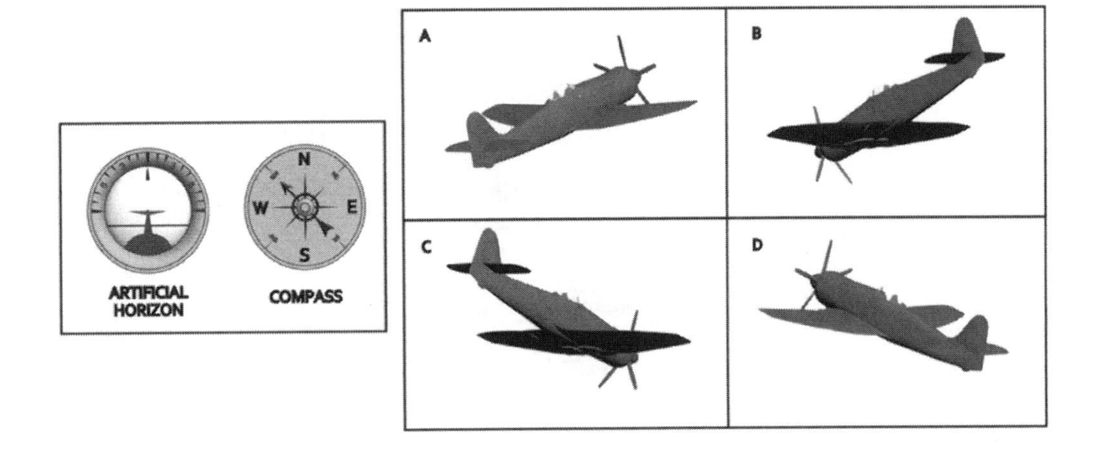

7. Which of the answer choices represents the orientation of the plane?

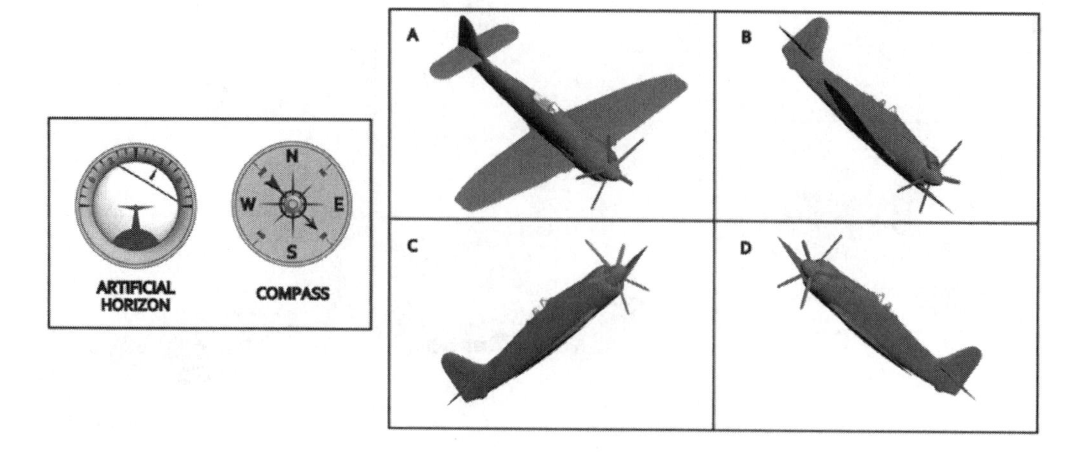

8. Which of the answer choices represents the orientation of the plane?

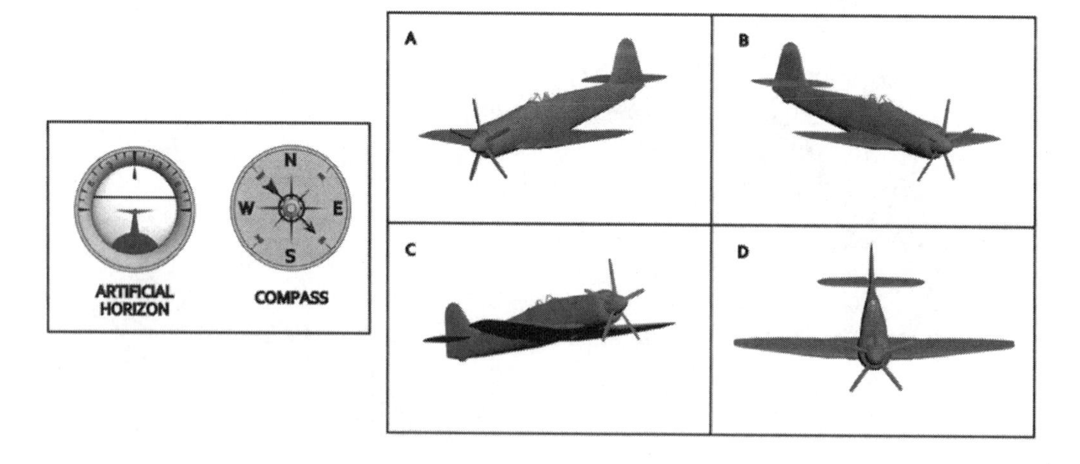

9. Which of the answer choices represents the orientation of the plane?

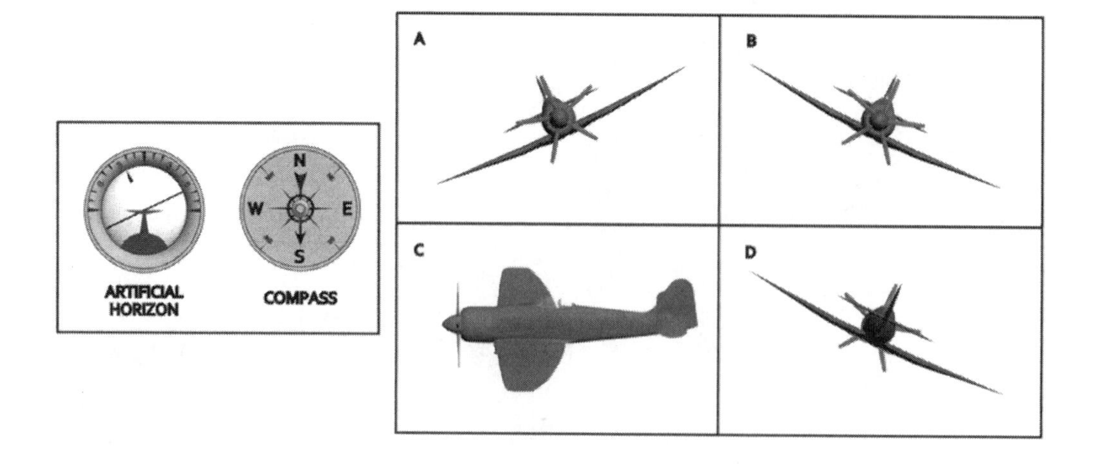

10. Which of the answer choices represents the orientation of the plane?

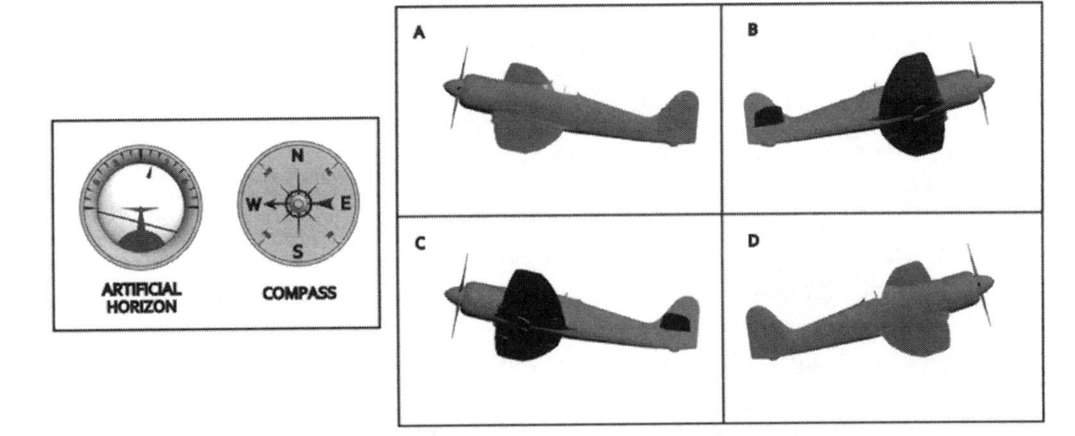

Practice Answers

1. The answer is D. The plane is flying east, which eliminates choices B and C. The artificial horizon indicates that the right wing of the plane is tilted upward. Only choice D meets this requirement.

2. The answer is C. The plane is facing west, which eliminates choice B. The artificial horizon indicates that the left wing of the plane is facing upward, so the answer is choice C.

3. The answer is A. The plane is facing north, which eliminates choices C and D. The artificial horizon indicates that the left wing is tilted upward, so the answer is choice A.

4. The answer is B because this is the only answer choice in which the plane is facing south.

5. The answer is C. The miniature wings indicate that the nose is tilted downward, which eliminates choices A and D. Of the two remaining choices, the answer must be C because the plane is flying southwest.

6. The answer is D. The artificial horizon indicates that the nose is tilted upward, which eliminates choices B and C. In addition, the compass is pointing northwest. Only choice D meets both requirements.

7. The answer is B. The compass shows that the plane is flying southeast, so you can eliminate choice D. In addition, the artificial horizon indicates that the nose of the plane is tilted downward, and the right wing of the plane is tilted upward. The only choice that meets both requirements is B.

8. The answer is B. The miniature wings in the artificial horizon are below the horizon bar, so the nose of the plane is tilted downward. As explained earlier, a plane that is traveling north is facing into the page. Therefore, a plane that is going southeast will have the orientation shown in choice B.

9. The answer is A. Because the compass is pointing south, the plane must appear to fly out of the page, so you can eliminate choices C and D. In addition, the artificial horizon indicates that the plane's left wing is tilted upward. Thus, B is the correct answer.

10. The answer is A. The compass indicates that the plane is flying west, which eliminates choices B and D. Also, the artificial horizon indicates that the right wing of the plane is tilted upward. Thus, A is the correct answer.

Block Counting

What are these questions testing?

These questions are designed to test your spatial, geometric, and logical abilities.

What is the question format?

The test will show a drawing of a three-dimensional arrangement of blocks with the same size and shape, and ask you to identify how many other blocks a particular block is touching. Typically, the blocks are arranged in irregular shapes, and some of the blocks are hidden. You will have to use spatial intuition and reasoning to determine how many blocks are touching the block in question.

How do I know how many blocks are touching the particular block?

First, it is important to know which blocks qualify as "touching" the other block. If at least part of a face of one block touches at least part of a face of another block, those blocks are considered to touch each other. However, if a block shares only a corner or an edge with another block, the two blocks do not touch.

The example below illustrates the difference between touching and non-touching blocks. Block A is touching blocks 1, 2, and 3 because part of a face of block A is touching each of these blocks. However, blocks 4 and 5 do *not* touch block A because they only contact block A at an edge; that is, they share no area with any faces of block A. Therefore, block A is touching three blocks in this picture.

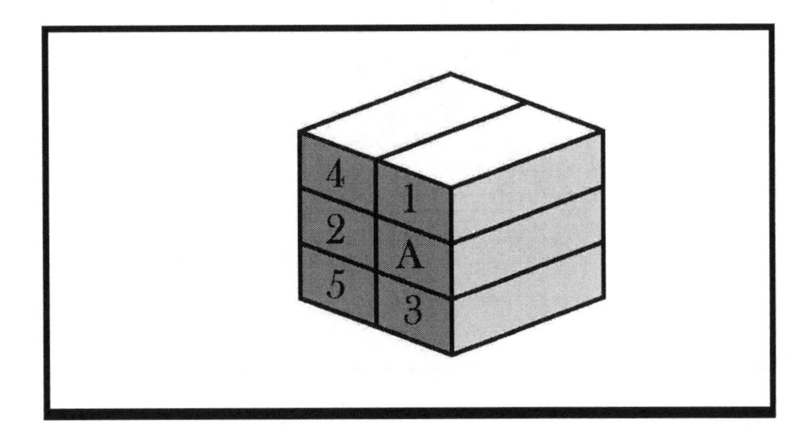

Sometimes, the blocks may be positioned so that certain blocks are hidden from view. In these cases, you will have to use basic spatial intuition and logic to determine the number of blocks touching a particular block.

In the example below, try to figure out how many blocks are touching block A.

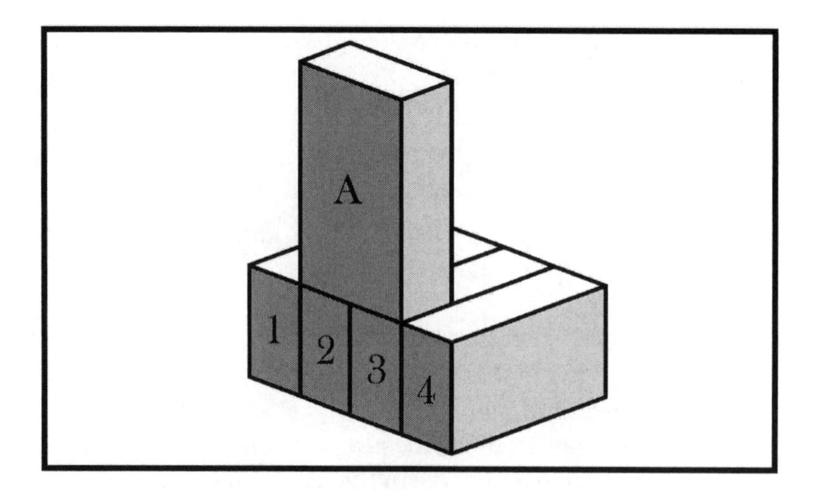

Answer: There are two blocks touching block A. Even though blocks 1 and 4 each share an edge with block A, they do not "touch" block A, as the word is used in the context of the test. Only blocks 2 and 3 actually touch a face of block A, so block A is only touching two blocks.

Are there any ways to make it easier to count the number of touching blocks?

Remember that for a block to count as "touching," it much contact a face of that block. Therefore, it might help to count the number of blocks that are touching each face of the block in question.

Try to apply this strategy to the example below, in order to count the number of blocks touching block A.

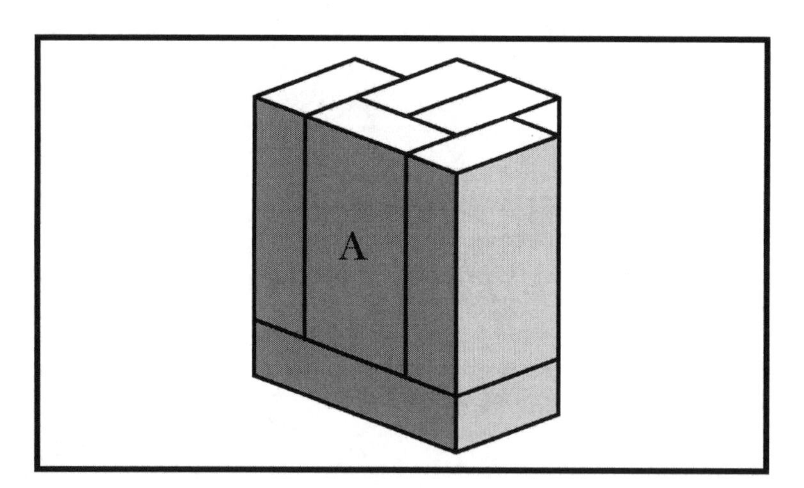

Answer: The back face of block A is touching two blocks. The left face of block A is touching one block, as is the right side. The bottom face is touching one block. Therefore, block A is touching a total of five blocks.

If you are still having trouble with the practice questions, it might help to try to re-create the example problems with a set of rectangular blocks. Practicing with physical blocks will make it easier to visualize hidden blocks when the blocks are drawn on paper.

Practice Questions

The explanations to these problems use terms like "top," "left," and "front" to refer to the faces of the blocks. Because this terminology can be confusing due to the angled view of the block arrangements, the illustration below is provided to help you keep these terms straight.

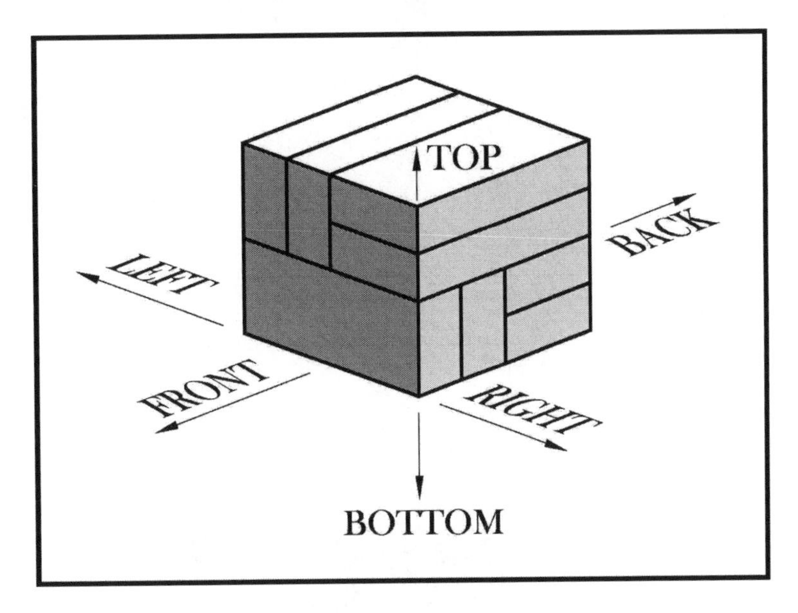

Once you understand these terms, proceed to the practice problems below.

1. How many blocks are touching block 1 in the arrangement below?

2. How many blocks are touching block 2 in the arrangement below?

3. How many blocks are touching block 8 in the arrangement below?

4. How many blocks are touching block 14 in the arrangement below?

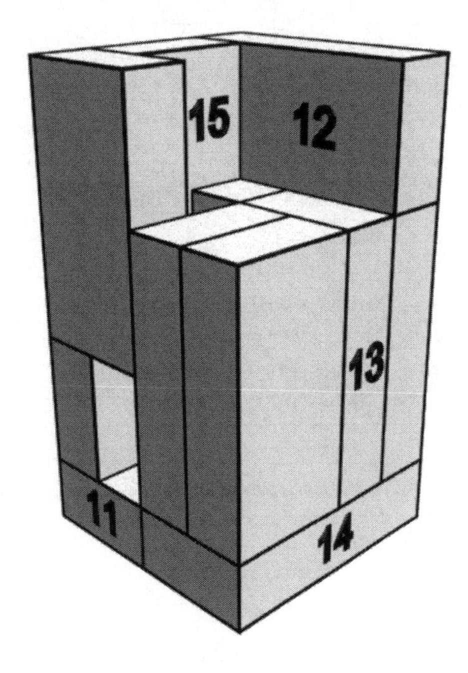

5. How many blocks are touching block 18 in the arrangement below?

Practice Answers

1. There are a total of six blocks touching block 1. One block (block #4) is touching the top face of block 1, two blocks are touching the left face, and three blocks are touching the bottom face.

2. There are a total of seven blocks touching block 2. Two blocks are touching the left face of block 2, one block (block 4) is touching the right face, and four blocks are touching the bottom face.

Note that the picture only directly shows two blocks (block 5 and the block in front of it) touching the bottom face of block 2. However, you can deduce from the picture that block 3 and the block in front of it must also touch block 2. Because the width of each block is twice the thickness, the width of block 3 and the block next to it must extend underneath block 5. Therefore, block 5 touches both of these blocks, even if it isn't explicitly shown in the picture.

3. A total of eight blocks are touching block 8. One block is touching the left face, one block is touching the right face, three blocks are touching the bottom face, one block is touching the top face, and two blocks are touching the back face.

4. A total of five blocks are touching block 14. One block (block 11) is touching the left face, and four blocks are touching the top face.

5. A total of nine blocks are touching block 18. Four blocks, including blocks 19 and 17, are touching the front face of block 18; four blocks are touching the back face; and one block is touching the bottom face.

Aviation Information

What are these questions testing?

The aviation information section tests your knowledge of basic aviation information. This includes a variety of things including aircraft terminology, the basic physics involved in flight, and common airport information.

How can I improve my ability to answer these questions?

Since these are all knowledge based questions, you can improve your success rate here by reading up on aircraft operation and airport information. The section below is an overview of the basics in these areas.

Fixed-wing aircraft

There are six basic components of a fixed-wing aircraft: wings; fuselage; tail assembly; landing gear; powerplant; and flight controls and control surfaces.

Wings
The **wings** are the primary airfoils of the plane. An airfoil is anything designed to produce lift when it moves through the air. The leading edge of an airfoil is thicker and rounder than the trailing edge, and the top surface of the airfoil has a greater curve than the bottom. The result is that air flows more quickly over the top of the wing, and the greater air pressure beneath pushes the wing, and thus the plane, upwards.

The wings connect to either side of the fuselage. Planes are designated as high-, mid-, and low-wing, depending on where the wings are attached. The wings themselves are described as either cantilever or semi-cantilever. A cantilever wing has sufficient internal support structures to keep it steady in its location. A semi-cantilever wing, on the other hand, requires additional external support structures. The trailing edge of a wing typically has two control surfaces attached by means of a hinge: flaps run from the fuselage to the middle of the wing, and ailerons run from the middle of the wing to the tip. By raising and lowering the ailerons, the pilot can roll the plane. The plane will roll when the ailerons are pointed in opposite directions. When the plane is cruising, however, these control surfaces are aligned with the rest of the wing. During takeoff and landing, both surfaces are extended, which increases the lift.

The distance from one wingtip to the other is the wingspan, and the distance from the leading edge to the trailing edge is called the chord. The chord line runs through the wing from leading edge to trailing edge: it divides the wing into upper and lower surfaces. The mean camber line runs along the inside of the wing, such that the parts of the wing above and below it are equal in thickness. The camber is the curvature of the airfoil: if an airfoil is heavily curved, it has a high camber. The thickness of a wing is measured at its greatest point. The shape of the wings when viewed from overhead is known as the planform.

When the wings are not attached parallel to the horizontal plane, the angle they make with the horizontal plane is called the dihedral angle. A positive dihedral wing angle (wings angling above the horizontal plane) keeps the plane stable when it rolls, as it will encourage the plane to return to its original position. This does diminish the maneuverability of the plane, which is why the wings of fighter jets are usually horizontal or even pointed slightly downwards (anhedral).

The shape of the wings has a major influence on the handling, maneuverability, and speed of the plane. Today's planes generally have a straight, sweep, or delta shape. Straight wings may be rectangular, elliptical (rounded), or tapered. They are commonly found on sailplanes, gliders, and other low-speed aircraft.

A swept wing provides better handling at high speeds, but makes the plane slightly less stable at low speeds. Since most modern aircraft are designed to operate at high speeds, this is the most commonly used wing style. Wings may be swept forward or back, though forward-swept wings are rarely seen. In general, a higher angle of sweep is used for planes that are meant to travel faster and be more maneuverable; however, more extreme sweeps require much greater speeds for takeoff and landing.

The delta wing shape is triangular, so that the leading edge of the wing has a high sweep angle while the trailing edge is mostly, if not completely, straight. A delta shape enables the plane to travel but also requires very high takeoff and landing speeds. Many of the earliest supersonic aircraft used the delta wing shape, as did the space shuttles.

Fuselage

The **fuselage** is the main body of the airplane. The basic features of the fuselage are the cockpit, cabin, cargo area, and attachment points for external components, like the wings and landing gear. Some planes designed for specific purposes may not have all of these components; for instance, a fighter jet will not have a cabin for passengers or a cargo area, since it needs to be light and maneuverable. The fuselage may be described as either truss or monocoque, depending on whether its strength is created by triangular arrangements of steel or aluminum tubing or by bulkheads, stringers and formers. A stringer is a support structure that runs the length of the fuselage, while a former runs perpendicular.

Tail assembly

The **tail assembly**, or empennage, includes the vertical and horizontal stabilizers, elevators, rudders, and trim tabs. The stabilizers are fixed (non-adjustable) surfaces that extend from the back end of the fuselage. The elevators are positioned along the trailing edges of the horizontal stabilizers; the pilot can move them to raise or lower the nose of the plane. The rudders are connected to the trailing edge of the vertical stabilizer, and are used to move the nose of the plane to the left or right, typically in combination with the ailerons. The trim tabs, finally, are movable surfaces that extend off the trailing edges of the rudder, elevators, and ailerons, and are used to make smaller adjustments.

Landing gear

The **landing gear** usually consists of three sets of wheels used for takeoffs and landings, though some planes have special non-wheel landing gear for landing on snow or water. Landing gear is commonly retractable, meaning that it is pulled up inside the plane during flight to reduce drag. In a typical arrangement, wheel sets are positioned either under each wing or on the sides of the fuselage, with the third wheel set being under the nose or the tail. Having the third wheel set under the nose, known as **tricycle** arrangement, is the most common arrangement on modern aircraft, but having the third wheel set under the tail is still known as **conventional** arrangement. Whether located under the tail or the nose, the third wheel will typically be able to rotate so that the plane can turn while traveling on the ground. The addition of extra wheels to each set allows the plane to handle a greater weight.

Powerplant

In aviation, the **powerplant** is the part of the plane that supplies the thrust. A jet engine operates by compressing the air that comes in the front, burning it along with fuel, and then blasting it out the back. There are different methods for compressing the air, but most jet engines do so by slowing it down with a set of small rotating blades. This greatly increases the air pressure at the front of the engine. The compressed air is then forced into a different section, where it is mixed with fuel and burned. As it then expands, it is pushed at great force through a series of turbines, the turning of which moves the compressor blades at the front of the engine, supplying both power and air. The exhausted air then passes out the back of the engine, which propels the plane forward. Some jet engines have afterburners, which feed extra fuel into the area between the turbines and the rear exhaust, increasing forward thrust.

In a propeller plane, on the other hand, the powerplant is the propellers and the engine. The propellers have tilted blades, which push air backwards and thereby push the plane forward. There are two types of propeller: fixed-pitch or variable-pitch. The blade angle of a fixed-pitch propeller cannot be adjusted by the pilot. Variable-pitch propellers allow the pilot, usually indirectly via the plane's control systems, to adjust the pitch of the propeller blades to alter the amount of thrust being generated. Some variable-pitch propellers are designed to operate only at a single rotational speed, allowing the engine to be much simpler and more efficient, so the amount of thrust is controlled entirely by the pitch of the blades. These are known as constant-speed propellers.

The engines of a propeller plane turn the crankshafts, which turn the propellers. The engines also are responsible for powering the plane's electrical system. The location of the engines on a propeller plane may vary. Single-engine planes typically have their engines in front of the fuselage, while multi-engine planes usually have their engines underneath the wings. Some multi-engine planes have engines in both locations.

Flight envelope

During flight, there are four forces a pilot must manage: lift, gravity, thrust, and drag. These forces act downward (gravity), upward (lift), forward (thrust), and backward (drag). The collective input of these forces is known as the flight envelope.

Gravity

The weight of a plane is the primary force that must be overcome for flight to take place. The force of **gravity** on a given object is the same, regardless of orientation, though it varies slightly with large changes in altitude.

Aviation experts distinguish between different types of weight. The basic weight includes the aircraft and any internal or external equipment that will remain a part of the plane during flight. The operating weight is the basic weight plus the crew and any other nonexpendable items not included in the basic weight. The gross weight is the total weight of the aircraft and all contents at any given time. The weight of the airplane when it has no usable fuel is called the zero fuel weight.

Lift

In order to overcome gravity, the plane must generate **lift**. Lift is the upward force of air pressure on the aircraft, primarily the wings, that allows it to achieve and maintain altitude. In order to generate lift, the plane typically must be traveling forward at considerable speed.

If the wing tilts too far back the airflow may stop over the wing's upper surface, which will result in a rapid loss of altitude and often control of the plane. This is known as a stall, and it may be avoided by decreasing the angle of attack, so that normal airflow over the top of the wing is not interrupted.

Thrust

The speed required for generating lift is provided by the aircraft's **thrust**. It ensures that the aircraft is able to continue moving forward at sufficient speed to generate lift. As was discussed in the previous section, thrust is generated by the powerplant of the aircraft, usually one or more jet engines or propellers.

Drag

An aircraft's thrust is countered by **drag**, the resistance to forward movement provided by the air that the aircraft is traveling through. At anything above normal walking or running speeds, air resistance is a noticeable hindrance to motion, and it only increases as airspeed goes up. The primary implication that this has on aircraft is that the faster the aircraft goes, the more thrust is required just to overcome the drag and maintain a constant speed.

There are two types of drag: profile drag and induced drag. Profile drag is the drag that exists when any object moves through the air. It is the result of the plane pushing air aside as it moves. Profile drag can be minimized by designing the aircraft to have a better wind profile. Induced drag, on the other hand, is drag that results from the wings generating lift. Part of the process of generating lift involves the wings redirecting the oncoming air downward (think Newton's third law), and this causes additional drag.

Atmospheric conditions

The flight envelope is significantly affected by the atmospheric conditions, primarily the density of the surrounding air and the speed and direction of any wind. The density is turn determined by the temperature, pressure, and humidity of the air. Lower temperatures, higher pressures, and lower humidity are all associated with higher density air. Denser air will produce greater lift, but will also produce more drag. Air pressure is most closely associated with altitude. In general, pressure decreases with altitude, so as you go higher up, the pressure of the surrounding air decreases.

With regard to wind, flying into the wind (headwind) has a similar type of impact to flying in denser air, though of much greater magnitude. In a headwind, the aircraft will have a higher speed relative to the surrounding air, which means it will experience greater drag and lift forces. Similarly, if the aircraft is flying the same direction as the wind (tailwind), it will have a lower speed relative to the surrounding air, and will experience reduced drag and lift.

Flight concepts and terminology

Flight attitude

The flight attitude is described in terms of three axes, all of which meet at the plane's center of mass.

The **longitudinal** axis is the axis that extends from the center forward toward the nose and rearward toward the tail. The **lateral** axis extends from the center out to the right and left, perpendicular to the longitudinal axis. Typically the lateral axis passes through (over/under) the wings. Both of these axes are in the horizontal plane when the aircraft is level. The **vertical** axis meanwhile extends straight upward and downward from the aircraft's center, perpendicular to the other two axes. The motion of the aircraft can be described in relation to these axes: Rotation about the longitudinal axis is called roll; rotation about the lateral axis is called pitch; rotation about the vertical axis is called yaw. In turn, these three types of motion are controlled by three sets of flight control surfaces. Roll is controlled by the ailerons, pitch by the elevators, and yaw by the rudder. This information is summarized in the table below, and is expanded upon in the following section.

Axis	Motion	Control surface
Longitudinal	Roll	Ailerons
Lateral	Pitch	Elevators
Vertical	Yaw	Rudder

Flight controls

Flight controls are divided into primary and secondary groups.

Primary

The primary **flight control surfaces** are the ailerons, rudder, and elevator.

The **ailerons** are responsible for the roll, or movement around the longitudinal axis. The ailerons extend from the trailing edges of the wings as shown in the figure below and can be manipulated by the pilot to cause the wing to either dip below or elevate above the horizontal plane.

The joystick (or control wheel) controls the roll of the aircraft. By pushing the stick (or turning the wheel) to the left, the pilot raises the left aileron and lowers the right aileron, causing the left wing to dip and the right wing to elevate.

The **elevators** control the plane's pitch, or movement around the lateral axis. They are attached to the trailing edges of the horizontal stabilizers at the rear of the aircraft. Depending on the design of the plane, there may be one elevator that extends across the length of the horizontal stabilizer, or there may be two elevators, divided by the vertical stabilizer, as shown in the figure below. When the elevators are undivided, they are sometimes referred to as a stabilator.

The joystick also controls the pitch of the aircraft. By pulling the stick back, the pilot raises the elevators, causing the tail of the plane to experience downward force, thus raising the nose of the plane. Pushing the stick forward will have the opposite effect on the elevators and will result in the nose of the plane dropping as the tail is pushed upward.

The **rudder** is a large flap attached by a hinge to the vertical stabilizer. It controls the motion of the plane around its vertical axis. The rudder can swing to the right or the left, causing the plane to turn (yaw) in either direction.

The rudder is controlled with two pedals: when the pilot pushes on the right pedal, the rudder swings out to the right, causing leftward pressure on the tail of the aircraft. This results in the nose of the plane turning to the right. Similarly, if the pilot pushes on the left pedal, the rudder will swing to the left, causing the tail to move right, and the nose to turn left.

The pilot also controls the amount of power or thrust being produced by the engines by manipulating the **throttle**. It is considered a primary flight control because the pilot must manage the thrust to ensure that the plane will be able to accomplish its intended maneuvers. With all three of the primary control surfaces, it is important to remember that the speed of the aircraft relative to the surrounding air determines the magnitude of the aircraft's response to the control. A plane that is traveling at 300 mph will roll much more quickly than one that is traveling at 200 mph in response to the same amount of aileron manipulation. The same is true of the other two types of motion.

Flight maneuvers usually involve the use of multiple controls. To make a proper turn, for instance, the pilot will need to employ the rudder, ailerons, and elevators. The bank is established by raising and lowering the ailerons, and the rudder pedals counteract any adverse yaw that occurs. Adverse yaw is the drifting of the nose caused by the extra drag on the downward-pointing aileron. Also, because extra lift is needed during a turn, the pilot must increase the angle of attack by applying downward elevator pressure. The amount of back elevator pressure required will be in proportion to the sharpness of the turn. This will be discussed in greater detail in the section on flight maneuvers.

Secondary

The secondary flight control surfaces include the flaps and leading edge devices, spoilers, and trim systems.

The **flaps** are connected to the trailing edges of the wings; they are raised or lowered to adjust the lift or drag. The retractable flaps on modern airplanes make it possible to cruise at a high speed and land at a low speed. On the opposite end of the wing, leading-edge devices accomplish much the same purpose. There are a number of different **leading-edge devices**: fixed slats, moveable slats, and leading edge flaps.

Spoilers are attached to the wings of some airplanes in order to diminish the lift and increase the drag. Spoilers can also be useful for roll control, in part because they reduce adverse yaw. This is accomplished by raising the spoiler on the side of the turn. This reduces the lift and creates more drag on that side, which causes that wing to drop and the plane to bank and yaw to that side. If both of the spoilers are raised at the same time, the plane can descend without increasing its speed. Raising the spoilers also improves the performance of the brakes, because they eliminate lift and push the plane down onto its wheels.

Trim systems exist mainly to ease the work of the pilot. They are attached to the trailing edges of one or more of the primary control surfaces. Small aircraft often have a single trim tab attached to the elevator. This tab is adjusted with a small wheel or crank, and its position is displayed in the cockpit. When the tab is deflected upwards, the trailing edge of the elevator is forced downward and the tail is pushed up, which lowers the nose of the plane.

Typically, a pilot first will achieve the desired pitch, power, attitude, and configuration, and then use the trim tabs to resolve the remaining control pressures. There are control pressures generated by any change in the flight condition, so trimming is necessary after any change. Trimming is complete when the pilot has eliminated any heaviness in the nose or tail of the plane.

Flight maneuvers

The four basic maneuvers in flight are straight-and-level flight, turning, climbing, and descending. As the name suggests, **straight-and-level flight** involves keeping the aircraft headed in a particular direction at a particular altitude. Maintaining straight-and-level flight requires frequent adjustment, much the same way as the driver of a car has to make frequent adjustments to maintain a straight path on a windy day or when driving on a rough uneven road.

Making a smooth **turn** requires the use of all four primary controls: the throttle is set to achieve a speed suitable to the desired type of turn, the ailerons bank the wings and the elevators raise the nose to establish the rate of turn, and the rudder is employed to counter any undesired yaw resulting from the effects of the other controls or to introduce desired yaw.

There are three classes of turn: shallow, medium, and deep. A shallow turn has a bank of less than 20 degrees. At angles this shallow, most planes will tend to try to stabilize themselves back to a level angle, so the pilot must maintain some pressure on the stick to ensure that the plane doesn't pull out of the bank prematurely. A medium turn has a bank of roughly 20 to 45 degrees. Most planes will tend to stay in a medium bank until the pilot makes an adjustment. Finally, a steep turn is one in which the bank is greater than 45 degrees. For angles this steep, most planes will tend to try to increase the banking angle even further unless the pilot counters that tendency by maintaining some pressure on the stick in a stabilizing direction.

While the pilot is getting the plane to the desired bank angle, he will also be pulling back on the stick to ensure that the nose of the plane does not dip during the bank. This also serves to increase the rate at which the plane turns its heading. In general, the steeper the bank, the more sharply the pilot must pull back on the stick to maintain altitude. Because the lowered aileron on the raised wing generally creates more drag than the raised aileron on the lowered wing, the airplane tends to yaw in the direction opposite to the turn. For this reason, the pilot must at the same time apply rudder pressure in the direction of the turn.

To initiate a **climb**, an aircraft's nose is angled upward so that it gains altitude. Several things that remain constant while the aircraft is flying level change when the nose of the plane is raised. The two most significant are the effective angle of gravity and the angle of attack of the wings.

When the aircraft is level, **gravity** acts entirely in direction of the vertical axis of the plane. When the plane angles upward, the force of gravity, which still acts straight down like before, now has a component in the longitudinal direction, since the rear of the plane is now pointed slightly toward the ground. Additionally, since raising the noise of the aircraft increases the angle of attack of the wings, the amount of **drag** the aircraft experiences goes up considerably during a climb. This means that, in order to maintain flight, the thrust must now overcome both an increased amount of drag and part of the force of gravity.

If the nose of the aircraft is raised to quickly or without a sufficient increase in thrust to account for the changing flight conditions, the aircraft may **stall**. The most common cause of a stall is the plane not generating enough thrust to maintain air speed, which means that the lift being generated by the wings is not sufficient to keep the plane in the air, so the plane ceases to fly in a practical sense and instead begins to fall. To correct a stall, the pilot must angle the nose of the aircraft steeply downward and increase the throttle to generate enough airspeed in the forward direction so that the control surfaces are effective in controlling the flight of the plane again, and pull out of the dive.

As should be apparent from this description, recovering from a stall involves significant loss of altitude, which makes stalling at low altitudes extremely dangerous.

There are a few different styles of controlled **descent**, but they all involve manipulation of the same two factors: **pitch** and **thrust**. By angling the nose of the plane downward, the pilot reduces the angle of attack of the wings and consequently reduces the amount of lift generated by the wings. This causes the plane to lose altitude. Similarly, by pulling back on the throttle, the pilot reduces the amount of thrust being generated, which in turn reduces the plane's air speed and the amount of lift generated by the wings, also resulting in a loss of altitude.

A **glide** is a controlled descent in which little or no engine power is used, and the plane drifts downward at a regular pace. The pilot manages a glide by balancing the forces of lift and gravity as they act on the plane.

When a pilot is executing a **landing**, the nose of the plane will actually be angled upward, but the throttle will be pulled way back to ensure that the plane continues its descent all the way to the ground.

Helicopters

In many ways, the operation of a helicopter is based on the same fundamentals as airplane flight. A helicopter is subject to the same four fundamental forces of lift, weight, thrust, and drag. Unlike an airplane, however, a helicopter applies most of its thrust vertically. When a helicopter flies at a constant speed in a stable horizontal path, the lift is equal to the weight and the forward thrust is equal to the drag. The helicopter will increase its horizontal speed if the thrust is greater than the drag, and will increase its altitude if the lift is greater than the weight. If the helicopter is hovering (i.e., not moving at all), there is no drag or forward thrust; only gravity and vertical thrust or lift, which are balanced.

The manner in which a helicopter generates lift is considerably different from that of an airplane. Whereas a plane derives its lift from the natural flow of air over the wing, the helicopter spins its "wing" rapidly and at a variable angle, giving it a variety of options for angles of attack. Because the main rotor of the helicopter is being torqued with such great force, it exerts the same amount of torque back on the fuselage of the helicopter but in the opposite direction (Newton's third law again). This necessitates a tail rotor to provide the force required to the keep the fuselage from spinning around while in flight. This function of the tail rotor is called **torque control**. Manipulation of the tail rotor is also used to change the heading of the helicopter.

Helicopter controls

Piloting a helicopter requires the use of three controls: the cyclic (stick), the collective, and the directional control system. The cyclic controls the longitudinal and lateral movement of the helicopter by adjusting the tilt of the main rotor. Moving the stick forward tilts the rotor forward, which in turn pushes the helicopter forward.

The collective is a tube running up from the cockpit floor to the left of the pilot. It has a handle that may be raised or lowered to affect the pitch, as well as a throttle that wraps around the handle and can be used to alter the engine torque. The collective controls the angle of the main rotor blades. If the handle is pulled up, the leading edge of the rotor blade lifts relative to the trailing edge.

The directional control system is a pair of pedals the pilot uses to alter the pitch of the tail rotor blades. Pressing one or the other of the pedals will cause the tail rotor to exert more or less force on the fuselage, which will in turn affect the heading of the helicopter.

A helicopter pilot must use all three of the controls at the same time. The cyclic and collective adjust the action of the main rotor, which must be compensated for with adjustment to the tail rotor. For instance, if the speed of the main rotor increases during a climb, the pilot will need to increase the amount of force generated by the tail rotor to ensure the fuselage does not begin to rotate.

If the helicopter loses engine power for some reason, the pilot will need to rely on autorotation, or the spinning of the rotors that is generated by airflow rather than the engine. The amount of torque on the fuselage will be smaller during autorotation, but it will still be enough to require the use of the tail rotor.

Unique forces

A helicopter generates some other forces that distinguish it from an airplane. **Translational lift** is extra lift a helicopter experiences when traveling in a forward direction.

The **Coriolis force** is another physical phenomenon related to helicopters. The Coriolis force is the increase in rotational speed that occurs when the weight of a spinning object moves closer to the rotation center. In the case of a helicopter, having a greater portion of the weight closer to the base of the blade will cause the rotor to move faster, or to require less power to move at the same rotational speed.

If the main rotor increases the flow of air over the rear part of the main rotor disc, than the rear part will have a smaller angle of attack. The result of this will be less lift in the rear part of the rotor disc. This is called the **transverse flow effect**. However, when a force is applied to a spinning disc, the effects will occur ninety degrees later. This phenomenon is known as gyroscopic precession.

Airport information

At an **airport**, the areas controlled by the aircraft traffic controller are called the movement (or maneuvering) areas. These include the runways and taxiways. Runways may be composed of all different materials, ranging from grass and dirt to asphalt and concrete. At a general aviation airport, the runways may be as little as 800 feet long and 26 feet wide, while an international airport may have runways that are 18,000 feet long and 260 feet across. The markings on a runway are white, but are usually outlined in black so that they may be better seen. Taxiways and areas not meant to be traveled by aircraft are marked in yellow.

There are three basic types of runway: visual, nonprecision instrument, and precision instrument. **Visual runways** are typical of small airports: they have no markings, though the boundaries and center lines may be indicated in some way. They are called visual runways because the pilot must be able to see the ground in order to land. It is not possible to land a plane on a visual runway simply with the use of instruments.

With a **nonprecision instrument runway**, a pilot may be able to make his approach using instruments. Specifically, this sort of runway can provide feedback on the horizontal position of the plane as it nears. Nonprecision instrument runways are commonly found at small and medium airports. These runways may have threshold markings, centerlines, and designators. These runways may also have a special mark, called an aiming point, between 1000 and 1500 feet long along the centerline of the runway.

Medium and large airports will have **precision instrument runways**, which give the pilot feedback on both horizontal and vertical position when the plane is on instrument approach. A precision instrument runway includes thresholds, designators, centerlines, aiming points, blast pads, stopways, and touchdown zone marks every 500 feet from the 500 foot to the 3000 foot mark.

Runways are named according to their **direction** on the compass, ranging from 01 to 36. So, for instance, due south would be runway 18 ("one-eight"), and due west would be runway 27 ("two-seven"). In North America, the runways are named in accordance with geographic (grid) north, rather than magnetic north. Of course, a runway may have two names, one for each direction in which it is used. The same runway may be referred to as runway 05 ("zero-five") or runway 23 ("two-three") depending on the direction it is being used on a given day. In most cases, fixed-wing aircraft take off and land against the wind, because the extra amount of air over the wing will increase lift (and reduce the required ground speed).

In the event that **multiple runways** travel in the same direction, they will be distinguished from each other by their relative positions according to an observer on approach from the appropriate direction: left or right runway if there are only two; left, right, or center runway if there are three. Of course, a runway that is on the right when travelling in one direction will be on the left when it is being used in the opposite direction.

In most cases, **runway lights** are operated by the airport control tower. There are a number of different components to a runway lighting system. A Runway Centerline Lighting System is a line of white lights mounted every fifty feet along the centerline. When the approaching plane gets within 3000 feet of the runway, the lights begin to blink red and white; when the plane gets within 1000 feet, the lights become solid red. Precision instrument runways have runway end lights and edge lights. Runway end lights run the width of both ends of the runway: from the ground these lights appear red, while they appear green from above. Runway edge lights run the length of the runway on both sides. This lighting typically changes color as well when the plane gets within a certain distance of the front end of the runway. There are similar lights marking the boundaries of taxiways. An Approach Lighting System is a set of strobelights and/or lightbars that indicate the end of the runway from which descending aircraft should arrive. Runway end identification lights are synchronized lights that flash at the runway thresholds. At some airports these lights face in every direction, while at others they only face the direction from which planes approach. Runway end identification lights are useful when the runway doesn't stand out from the surrounding area, or when visibility is poor.

Some big airports also have **Visual Approach Slope Indicators**, which give the incoming pilot useful information. In a typical VASI system, white lights indicate the lower glide path limits and red lights indicate the upper. The VASI should be visible for twenty miles at night, and for three to five miles during the day under normal conditions. An effective VASI should keep the plane clear of obstructions so long as it remains within approximately ten degrees of the extended runway centerline and within four nautical miles of the runway threshold.

Practice Questions

1. What would be the name of a runway that the pilot approaches while heading due east?
    a. Runway 01
    b. Runway 09
    c. Runway 27
    d. Runway E
    e. Runway B

2. Which part of a fixed-wing airplane supplies the thrust?
    a. tail assembly
    b. control surfaces
    c. fuselage
    d. landing gear
    e. powerplant

3. What is the path of the chord line on a wing?
    a. from leading edge to trailing edge, through the wing
    b. from leading edge to trailing edge, along the surface of the wing
    c. along the inside of the wing, such that the upper and lower wings are equal in thickness
    d. from one wingtip to the other
    e. from wingtip to fuselage

4. Which of the following statements about a medium bank is true?
    a. A plane will tend to level out from a medium bank unless there is input from the pilot.
    b. A plane that is put into a medium bank will tend to increase its bank unless the ailerons are applied.
    c. A medium bank is between ten and thirty degrees.
    d. A medium bank is between thirty and fifty degrees.
    e. A plane will tend to remain in a medium bank until the pilot makes an adjustment.

5. Which of the following is NOT one of the four basic maneuvers in flight?
    a. turn
    b. spin
    c. straight-and-level flight
    d. climb
    e. descent

Practice Answers

1. B: The name of a runway pointing due east would be Runway 9. The names of runways are based on the compass. A runway pointing due south would be Runway 18, and a runway pointing due west would be Runway 27. Of course, every runway will be called by two names, depending on which direction planes are traveling on it. Runway 9 will become Runway 27 when the planes travel west rather than east. When the names of runways are spoken, each number in the name is stated individually. So, Runway 24 would be spoken, "Runway Two-Four" rather than "Runway Twenty-four."

2. E: The powerplant supplies the thrust for a fixed-wing aircraft. The powerplant may be a jet engine or an engine and a set of propellers. Thrust is the force that propels the plane forward.

3. A: The chord line runs through the wing from the leading edge to the trailing edge. This line divides the upper and lower surfaces of the wing. This is one of the key elements of wing design. The distance from one wingtip to the other, meanwhile, is called the wingspan. The mean camber line runs along the inside of the wing, such that the upper and lower wings are equal in thickness.

4. E: A plane will tend to remain in a medium bank until the pilot makes an adjustment. A medium bank is between 20 and 45 degrees. A shallow bank, on the other hand, is less than 20 degrees, and requires the assistance of the ailerons to maintain itself. A steep bank is greater than 45 degrees. When a plane enters a steep bank, it will tend to increase the bank unless the ailerons are used to prevent this. Turning occurs because of the forces that act on a banked wing. The plane will be pushed in a direction perpendicular to the wings.

5. B: Spin is not one of the four basic maneuvers in flight. Straight-and-level flight, turns, climbs, and descents are the four basic flight maneuvers. Straight-and-level flight occurs when the plane maintains a constant altitude and is pointed in the same direction. Of course, maintaining the same altitude and direction requires a number of adjustments. Turns are made by banking the wings in the direction of the turn: that is, for example, turning to the right requires lowering the right wing. A climb requires raising the nose of the plane and increasing the power from the engine. Finally, there are a few types of descent. A plane may descend with its nose up, down, or level.

Self Description Inventory

The final section of the AFOQT is a self description inventory, or basically a personality test. You will be shown a series of statements and you will have to decide how well those statements describe you. There are no right or wrong answers to the questions in this section, so don't spend a lot of thinking about them. Your first instinct will usually be the most accurate assessment of yourself.

AFOQT Practice Test

Verbal Analogies

1. CHASTISE is to REPRIMAND as IMPETUOUS is to
   a. punish
   b. rash
   c. considered
   d. poor
   e. calm

2. ARM is to HUMERUS as LEG is to
   a. ulna
   b. clavicle
   c. femur
   d. mandible
   e. metacarpal

3. MULTIPLICATION is to DIVISION as PRODUCT is to
   a. quotient
   b. divisor
   c. integer
   d. dividend
   e. multiplier

4. WEAR is to SWEATER as EAT is to
   a. shirt
   b. top hat
   c. asparagus
   d. looks
   e. mouth

5. MONEY is to IMPECUNIOUS as FOOD is to
   a. famished
   b. nauseated
   c. distracted
   d. antagonistic
   e. impoverished

6. DENIGRATE is to MALIGN as DEMUR is to
   a. protest
   b. defer
   c. slander
   d. benumb
   e. belittle

7. OBEISANCE is to DEFERENCE as MUNIFICENT is to
 a. benevolent
 b. magnificent
 c. squalid
 d. generous
 e. avarice

8. GOAT is to NANNY as PIG is to
 a. shoat
 b. ewe
 c. cub
 d. sow
 e. calf

9. CACHE is to RESERVE as DEARTH is to
 a. stockpile
 b. paucity
 c. cudgel
 d. dirge
 e. somber

10. ARABLE is to FARMABLE as ASYLUM is to
 a. famine
 b. danger
 c. arid
 d. fertile
 e. refuge

11. MYRIAD is to FEW as STATIONARY is to
 a. peripatetic
 b. many
 c. several
 d. halted
 e. parked

12. HOUSE is to MANSION as BOTTLE is to
 a. flagon
 b. container
 c. vessel
 d. pot
 e. flask

13. DICTIONARY is to DEFINITIONS as THESAURUS is to
 a. pronunciations
 b. synonyms
 c. explanations
 d. pronouns
 e. definitions

14. ABSTRUSE is to ESOTERIC as ADAMANT is to
    a. yielding
    b. stubborn
    c. keen
    d. forthright
    e. flexible

15. BEES is to HIVE as CATTLE is to
    a. swarm
    b. pod
    c. herd
    d. flock
    e. pack

16. LATITUDE is to LONGITUDE as PARALLEL is to
    a. strait
    b. line
    c. equator
    d. aquifer
    e. meridian

17. PREVENTION is to DETERRENCE as INCITEMENT is to
    a. excitement
    b. provocation
    c. request
    d. disregard
    e. disgust

18. VALUE is to WORTH as MEASURE is to
    a. gauge
    b. allowance
    c. demerit
    d. insignificance
    e. large

19. ENERVATE is to ENERGIZE as ESPOUSE is to
    a. oppose
    b. wed
    c. equine
    d. epistolary
    e. marry

20. HAMMER is to CARPENTER as STETHOSCOPE is to
    a. patient
    b. hearing
    c. heartbeat
    d. doctor
    e. pedometer

21. ARMOIRE is to BEDROOM as DESK is to
    a. chair
    b. office
    c. computer
    d. work
    e. building

22. PLANKTON is to WHALES as BAMBOO is to
    a. predators
    b. grasses
    c. pandas
    d. fast-growing
    e. China

23. TUNDRA is to ARCTIC as SAVANNA is to
    a. prairie
    b. lush
    c. tropic
    d. Georgia
    e. jungle

24. THIRSTY is to PARCHED as HUNGRY is to
    a. famished
    b. fed
    c. satiated
    d. satisfied
    e. full

25. FELICITY is to SADNESS as IGNOMINY is to
    a. shame
    b. slander
    c. crime
    d. indict
    e. honor

Arithmetic Reasoning

1. A man buys two shirts. One is $7.50 and the other is $3.00. A 6% tax is added to his total. What is his total?
   a. $10.50
   b. $11.13
   c. $14.58
   d. $16.80
   e. $18.21

2. If a chef can make 25 pastries in a day, how many can he make in a week?
   a. 32
   b. 74
   c. 126
   d. 175
   e. 250

3. A woman must earn $250 in the next four days to pay a traffic ticket. How much will she have to earn each day?
   a. $45.50
   b. $62.50
   c. $75.50
   d. $100.50
   e. $125.00

4. A car lot has an inventory of 476 cars. If 36 people bought cars in the week after the inventory was taken, how many cars will remain in inventory at the end of that week?
   a. 440
   b. 476
   c. 484
   d. 512
   e. 536

5. A woman has $450 in a bank account. She earns 0.5% interest on her end-of-month balance. How much interest will she earn for the month?
   a. $0.50
   b. $2.25
   c. $4.28
   d. $4.73
   e. $6.34

6. Three children decide to buy a gift for their father. The gift costs $78. One child contributes $24. The second contributes $15 less than the first. How much will the third child have to contribute?
   a. $15
   b. $39
   c. $45
   d. $62
   e. $69

7. Two women have credit cards. One earns 3 points for every dollar she spends. The other earns 6 points for every dollar she spends. If they each spend $5.00, how many combined total points will they earn?

a. 15
b. 30
c. 45
d. 60
e. 75

8. A company employing 540 individuals plans to increase its workforce by 13%. How many people will the company employ after the expansion?

a. 527
b. 547
c. 553
d. 570
e. 610

9. A 13 story building has 65 apartments. If each floor has an equal number of apartments, how many apartments are on each floor?

a. 2
b. 3
c. 4
d. 5
e. 6

10. If 5 people buy 3 pens each and 3 people buy 7 pencils each, what is the ratio of the total number of pens sold to the total number of pencils sold?

a. 15:21
b. 3:7
c. 5:7
d. 1:1
e. 5:3

11. A man earns $15.23 per hour and gets a raise of $2.34 per hour. What is his new hourly rate of pay?

a. $12.89
b. $15.46
c. $17.57
d. $23.40
e. $35.64

12. How many people can travel on 6 planes if each carries 300 passengers?

a. 1800
b. 1200
c. 600
d. 350
e. 300

13. In a town, the ratio of men to women is 2:1. If the number of women in the town is doubled, what will be the new ratio of men to women?
a. 1:2
b. 1:1
c. 2:1
d. 3:1
e. 4:1

14. A woman weighing 250 pounds goes on a diet. During the first week, she loses 3% of her body weight. During the second week, she loses 2%. At the end of the second week, how many pounds has she lost?
a. 12.5
b. 10
c. 12.35
d. 15
e. 17.5

15. A woman is traveling to a destination 583 km away. If she drives 78 km every hour, how many hours will it take for her to reach her destination?
a. 2.22
b. 3.77
c. 5.11
d. 7.47
e. 8.32

16. If one gallon of paint can paint 3 rooms, how many rooms can be painted with 28 gallons of paint?
a. 10
b. 25
c. 56
d. 84
e. 92

17. Five workers earn $135/day. What is the total amount earned by the five workers?
a. $675
b. $700
c. $725
d. $750
e. $775

18. A girl scores a 99 on her math test. On her second test, her score drops by 15. On the third test, she scores 5 points higher than she did on her second. What was the girl's score on the third test?
a. 79
b. 84
c. 89
d. 99
e. 104

19. A man goes to the mall with $50.00. He spends $15.64 in one store and $7.12 in a second store. How much does he have left?

a. $27.24
b. $32.76
c. $34.36
d. $42.8
e. $57.12

20. 600 students must share a school that has 20 classrooms. How many students will each classroom contain if there are an equal number of students in each class?

a. 10
b. 15
c. 20
d. 25
e. 30

21. Four workers at a shelter agree to care for the dogs over a holiday. If there are 48 dogs, how many must each worker look after?

a. 8
b. 10
c. 12
d. 14
e. 16

22. One worker has an office that is 20 feet long. Another has an office that is 6 feet longer. What is the combined length of both offices?

a. 26 feet
b. 36 feet
c. 46 feet
d. 56 feet
e. 66 feet

23. Four friends go shopping. They purchase items that cost $6.66 and $159.23. If they split the cost evenly, how much will each friend have to pay?

a. $26.64
b. $39.81
c. $41.47
d. $55.30
e. $82.95

24. A 140 acre forest is cut in half to make way for development. What is the size of the new forest's acreage?

a. 70
b. 80
c. 90
d. 100
e. 120

25. A farmer has 360 cows. He decides to sell 45. Shortly after, he purchases 85 more cows. How many cows does he have?

a. 230
b. 315
c. 400
d. 490
e. 530

Word Knowledge

1. The word **spoiled** most nearly means
    a. ruined
    b. splendid
    c. told
    d. believed
    e. hated

2. He made an **oath** to his king.
    a. delivery
    b. promise
    c. statement
    d. criticism
    e. threat

3. **Inquire** most nearly means
    a. invest
    b. ask
    c. tell
    d. release
    e. inquest

4. Spanish is a difficult language to **comprehend**.
    a. learn
    b. speak
    c. understand
    d. appreciate
    e. commemorate

5. **Apparent** most nearly means
    a. clear
    b. occasional
    c. angry
    d. applied
    e. father

6. They enjoyed the **silence** of the night.
    a. darkness
    b. excitement
    c. quiet
    d. mood
    e. quaint

7. **Absolutely** most nearly means
   a. assuredly
   b. rapidly
   c. never
   d. weakly
   e. completely

8. He **modified** his schedule so he could attend the staff lunch.
   a. checked
   b. shortened
   c. considered
   d. changed
   e. lengthened

9. **Delicate** most nearly means
   a. fragile
   b. sturdy
   c. loud
   d. soft
   e. lovely

10. She attended the New Year's **festivities**.
   a. commitments
   b. celebrations
   c. crowds
   d. dates
   e. funeral

11. **Exhausted** most nearly means
   a. excited
   b. tired
   c. worried
   d. energized
   e. animated

12. She **cleansed** her face in the morning.
   a. examined
   b. washed
   c. touched
   d. dried
   e. motivated

13. **Battled** most nearly means
   a. fought
   b. attempt
   c. bold
   d. saw
   e. excited

14. He **wandered** around the mall.
   a. looked
   b. shopped
   c. roamed
   d. searched
   e. lived

15. **Abruptly** most nearly means
   a. homely
   b. commonly
   c. wisely
   d. ugly
   e. suddenly

16. He was **tricked** into giving her money.
   a. conned
   b. begged
   c. convinced
   d. nagged
   e. criticized

17. **Extremely** most nearly means
   a. almost
   b. slightly
   c. very
   d. clearly
   e. happily

18. She was **doubtful** whether the plan would work.
   a. uncertain
   b. panicked
   c. pondering
   d. indifferent
   e. confused

19. **Peculiar** most nearly means
    a. original
    b. novel
    c. dull
    d. strange
    e. awesome

20. He is a very **courteous** young man.
    a. handsome
    b. polite
    c. inconsiderate
    d. odd
    e. unrelenting

21. **Troubled** most nearly means
    a. relieved
    b. satisfied
    c. bothered
    d. relaxed
    e. persistent

22. **Perspiration** most nearly means
    a. sweat
    b. work
    c. help
    d. advice
    e. job

23. The child **trembled** with fear.
    a. spoke
    b. shook
    c. wept
    d. ducked
    e. cowered

24. **Adhered** most nearly means
    a. stuck
    b. went
    c. spoke
    d. altered
    e. stunk

25. She kept her house **tidy**.
   a. furnished
   b. warm
   c. locked
   d. neat
   e. inviting

Math Knowledge

1. A rectangle has a width of 7 cm and a length of 9 cm. What is its perimeter?
   a. 16 cm
   b. 32 cm
   c. 48 cm
   d. 62 cm
   e . 63 cm

2. In the following inequality, solve for q.
$-3q + 12 \geq 4q - 30$
   a. $q \geq 6$
   b. $q = 6$
   c. $q \neq 6$
   d. $q \leq 6$
   e. q does not exist

3. If $x - 6 = 0$, then x is equal to
   a. 0
   b. 3
   c. 6
   d. 9
   e. 12

4. If $x = -3$, calculate the value of the following expression:
$3x^3 + (3x + 4) - 2x^2$
   a. -104
   b. -58
   c. 58
   d. 104
   e. 0

5. If $3x - 30 = 45 - 2x$, what is the value of x?
   a. 5
   b. 10
   c. 15
   d. 20
   e. 25

6. Solve for x in the following inequality.
$\frac{1}{4}x - 25 \geq 75$
   a. $x \geq 400$
   b. $x \leq 400$
   c. $x \geq 25$
   d. $x \leq 25$
   e. $x \geq 50$

7. If $x^2 - 5 = 20$, what is possible value of x?

a. 5
b. 10
c. 12.5
d. 15
e. 25

8. What is the area of a square that has a perimeter of 8 cm?

a. 2 cm^2
b. 4 cm^2
c. 32 cm^2
d. 64 cm^2
e. 160 cm^2

9. If x = 4 and y = 2, what is the value of the following expression:
$3xy - 12y + 5x$

a. -4
b. 10
c. 12
d. 20
e. 24

10. If $0.65x + 10 = 15$, what is the value of x?

a. 4.92
b. 5.78
c. 6.45
d. 7.69
e. 8.12

11. Simplify the following:
$(3x + 5)(4x - 6)$

a. $12x^2 - 38x - 30$
b. $12x^2 + 2x - 30$
c. $12x^2 - 2x - 1$
d. $12x^2 + 2x + 30$
e. $12x^2 + 7x - 30$

12. Simplify the following expression:

$$\frac{50x^{18}t^6w^3z^{20}}{5x^5t^2w^2z^{19}}$$

a. $10x^{13}t^3wz$
b. $10x^{13}t^4wz$
c. $10x^{12}t^4wz$
d. $10x^{13}t^4wz^2$
e. $10x^{12}t^3w^2z^2$

13. $4! =$
  a. 4
  b. 12
  c. 16
  d. 20
  e. 24

14. If a cube is 5 cm long, what is the volume of the cube?
  a. 15 cm^3
  b. 65 cm^3
  c. 105 cm^3
  d. 125 cm^3
  e. 225 cm^3

15. Solve for x by factoring:
$x^2 - 13x + 42 = 0$
  a. $x = 6, 7$
  b. $x = -6, -7$
  c. $x = 6, -7$
  d. $x = -6, 7$
  e. $x = 7$ only

16. A triangle has a base measuring 12 cm and a height of 12 cm. What is its area?
  a. 24 cm^2
  b. 56 cm^2
  c. 72 cm^2
  d. 144 cm^2
  e. 288 cm^2

17. Simplify the following expression:
$(3x^2 * 7x^7) + (2y^3 * 9y^{12})$
  a. $21x^{14} + 18y^{26}$
  b. $10x^9 + 11y^{15}$
  c. $21x^{14} + 18y^{15}$
  d. $21x^9 + 18y^{15}$
  e. $10x^{14} + 11y^{26}$

18. If $x/3 + 27 = 30$, what is the value of x?
  a. 3
  b. 6
  c. 9
  d. 12
  e. 27

19. What is the slope of a line with points A (4,1) and B (-13,8)?
  a. 7/17
  b. -7/17
  c. -17/7
  d. 17/7
  e. none of the above

20. If x is 20% of 200, what is the value of x?
a. 40
b. 80
c. 100
d. 150
e. 180

21. If a bag of balloons consists of 47 white balloons, 5 yellow balloons, and 10 black balloons, what is the probability that a balloon chosen randomly from the bag will be black?
a. 19%
b. 16%
c. 21%
d. 33%
e. 10%

22. In a lottery game, there are 2 winners for every 100 tickets sold. If a man buys 10 tickets, what are the chances that he is a winner?
a. 1 in 2
b. 1 in 5
c. 2 in 5
d. 2 in 2
e. 1 in 50

23. What is the volume of a rectangular prism with a height of 10cm, a length of 5cm, and a width of 6cm?
a. 30 cm^3
b. 60 cm^3
c. 150 cm^3
d. 240 cm^3
e. 300 cm^3

24. What is the midpoint of point A (6, 20) and point B (10, 40)?
a. (30, 8)
b. (16, 60)
c. (8, 30)
d. (7, 15)
e. (15, 8)

25. If 5x + 60 = 75, what is the value of x?
a. 3
b. 4
c. 5
d. 6
e. 7

Reading Comprehension

Passage 1

Since air is a gas, it can be compressed or expanded. When air is compressed, a greater amount of air can occupy a given volume. Conversely, when pressure on a given volume of air is decreased, the air expands and occupies a greater space. That is, the original column of air at a lower pressure contains a smaller mass of air. In other words, the density is decreased. In fact, density is directly proportional to pressure. If the pressure is doubled, the density is doubled, and if the pressure is lowered, so is the density. This statement is true only at a constant temperature.

Increasing the temperature of a substance decreases its density. Conversely, decreasing the temperature increases the density. Thus, the density of air varies inversely with temperature. This statement is true only at a constant pressure. In the atmosphere, both temperature and pressure decrease with altitude, and have conflicting effects upon density. However, the fairly rapid drop in pressure as altitude is increased usually has the dominant effect. Hence, pilots can expect the density to decrease with altitude.

The preceding paragraphs are based on the presupposition of perfectly dry air. In reality, it is never completely dry. The small amount of water vapor suspended in the atmosphere may be negligible under certain conditions, but in other conditions humidity may become an important factor in the performance of an aircraft. Water vapor is lighter than air; consequently, moist air is lighter than dry air. Therefore, as the water content of the air increases, the air becomes less dense, increasing density altitude and decreasing performance. It is lightest or least dense when, in a given set of conditions, it contains the maximum amount of water vapor.

Humidity, also called relative humidity, refers to the amount of water vapor contained in the atmosphere, and is expressed as a percentage of the maximum amount of water vapor the air can hold. This amount varies with the temperature; warm air can hold more water vapor, while colder air can hold less. Perfectly dry air that contains no water vapor has a relative humidity of zero percent, while saturated air that cannot hold any more water vapor has a relative humidity of 100 percent. Humidity alone is usually not considered an essential factor in calculating density altitude and aircraft performance; however, it does contribute.

The higher the temperature, the greater amount of water vapor that the air can hold. When comparing two separate air masses, the first warm and moist (both qualities making air lighter) and the second cold and dry (both qualities making it heavier), the first must be less dense than the second. Pressure, temperature, and humidity have a great influence on aircraft performance because of their effect upon density. There is no rule-of-thumb or chart used to compute the effects of humidity on density altitude, but it must be taken into consideration. Expect a decrease in overall performance in high humidity conditions.

1. The primary purpose of the passage is to
   a. explain the qualities of air that may affect flight.
   b. explain g-force and how it works.
   c. describe the constituent elements of air.
   d. explain humidity.
   e. describe the ideal air conditions for flight.

2. In the second paragraph, *inversely* most nearly means
   a. severely
   b. incredibly
   c. in the opposite direction
   d. in an unrelated fashion
   e. concurrently

3. If the air temperature drops while a plane is gaining altitude, the pilot can expect
   a. the density of the air to increase.
   b. the humidity of the air to increase.
   c. the air pressure to increase.
   d. the density of the air to decrease.
   e. aircraft performance to decrease.

4. With which one of the following claims about air quality would the author most likely agree?
   a. Pilots never need to pay attention to relative humidity.
   b. For a pilot, the density of air is more important than the relative humidity.
   c. Completely dry air is very rare.
   d. Aircraft performance is unrelated to humidity.
   e. The best conditions for flying are very hot and humid.

5. What is the most likely reason why there is no chart for assessing the effects of humidity on density altitude?
   a. Humidity does not affect density altitude.
   b. It is impossible to measure humidity.
   c. Humidity does not affect flight performance very much.
   d. Humidity varies a great deal in relatively small areas.
   e. Density altitude never varies.

Passage 2
The climb performance of an aircraft is affected by certain variables. The conditions of the aircraft's maximum climb angle or maximum climb rate occur at specific speeds, and variations in speed will produce variations in climb performance. There is sufficient latitude in most aircraft that small variations in speed from the optimum do not produce large changes in climb performance, and certain operational considerations may require speeds slightly different from the optimum. Of course, climb performance would be most critical with high gross weight, at high altitude, in obstructed takeoff areas, or during malfunction of a powerplant. Then, optimum climb speeds are necessary.

Weight has a very pronounced effect on aircraft performance. If weight is added to an aircraft, it must fly at a higher angle of attack (AOA) to maintain a given altitude and speed. This increases the induced drag of the wings, as well as the parasite drag of the aircraft. Increased drag means that additional thrust is needed to overcome it, which in turn means that less reserve thrust is available for climbing. Aircraft designers go to great effort to minimize the weight since it has such a marked effect on the factors pertaining to performance.

A change in an aircraft's weight produces a twofold effect on climb performance. First, a change in weight will change the drag and the power required. This alters the reserve power available, which in turn, affects both the climb angle and the climb rate. Secondly, an increase in weight will reduce the maximum rate of climb, but the aircraft must be operated at a higher climb speed to achieve the smaller peak climb rate.

An increase in altitude also will increase the power required and decrease the power available. Therefore, the climb performance of an aircraft diminishes with altitude. The speeds for maximum rate of climb, maximum angle of climb, and maximum and minimum level flight airspeeds vary with altitude. As altitude is increased, these various speeds finally converge at the absolute ceiling of the aircraft. At the absolute ceiling, there is no excess of power and only one speed will allow steady, level flight. Consequently, the absolute ceiling of an aircraft produces zero rate of climb. The service ceiling is the altitude at which the aircraft is unable to climb at a rate greater than 100 feet per minute (fpm). Usually, these specific performance reference points are provided for the aircraft at a specific design configuration.

In discussing performance, it frequently is convenient to use the terms power loading, wing loading, blade loading, and disk loading. Power loading is expressed in pounds per horsepower and is obtained by dividing the total weight of the aircraft by the rated horsepower of the engine. It is a significant factor in an aircraft's takeoff and climb capabilities. Wing loading is expressed in pounds per square foot and is obtained by dividing the total weight of an airplane in pounds by the wing area (including ailerons) in square feet. It is the airplane's wing loading that determines the landing speed. Blade loading is expressed in pounds per square foot and is obtained by dividing the total weight of a helicopter by the area of the rotor blades. Blade loading is not to be confused with disk loading, which is the total weight of a helicopter divided by the area of the disk swept by the rotor blades.

6. Which of the following would be the best title for this passage?
   a. The Importance of Weight
   b. Climb Performance and You
   c. Power Loading, Wing Loading, and Disk Loading
   d. Influences on Climb Performance
   e. Achieving Maximum Climb Angle

7. In the second paragraph, *pronounced* most nearly means
   a. selective
   b. intoned
   c. detrimental
   d. spoken
   e. noticeable

8. Which of the following is NOT one of the effects of increased weight on flight performance?
   a. diminished reserve power
   b. decreased climb rate
   c. lower angle of attack required to maintain altitude
   d. diminished maximum rate of climb
   e. increased drag

9. With which one of the following claims about climb performance would the author most likely agree?
   a. Optimal climb performance can be achieved even with heavy cargo.
   b. At the end of a long journey, a plane will have a higher maximum rate of climb.
   c. A plane can handle any amount of weight, though climb performance will be affected.
   d. Pilots have no influence over climb performance.
   e. The climb performance of a two-engine plane will remain the same even if one engine fails.

10. If a helicopter weighs two tons and its rotor blades cover an area of five hundred square feet, what is its disc loading measure?
   a. 4 pounds per square foot
   b. 250 square foot-pounds
   c. 0.25 tons per square foot
   d. 4 metric tons
   e. 3.65 ton-feet

Passage 3
The aerodynamic properties of an aircraft generally determine the power requirements at various conditions of flight, while the powerplant capabilities generally determine the power available at various conditions of flight. When an aircraft is in steady, level flight, a condition of equilibrium must prevail. An unaccelerated condition of flight is achieved when lift equals weight, and the powerplant is set for thrust equal to drag. The power required to achieve equilibrium in constant-altitude flight at various airspeeds is depicted on a power required curve. The power required curve illustrates the fact that at low airspeeds near the stall or minimum controllable airspeed, the power setting required for steady, level flight is quite high.

Flight in the region of normal command means that while holding a constant altitude, a higher airspeed requires a higher power setting and a lower airspeed requires a lower power setting. The majority of aircraft flying (climb, cruise, and maneuvers) is conducted in the region of normal command.

Flight in the region of reversed command means flight in which a higher airspeed requires a lower power setting and a lower airspeed requires a higher power setting to hold altitude. It does not imply that a decrease in power will produce lower airspeed. The region of reversed command is encountered in the low speed phases of flight. Flight speeds below the speed for maximum endurance (lowest point on the power curve) require higher power settings with a decrease in airspeed. Since the need to increase the required power setting with decreased speed is contrary to the normal command of flight, the regime of flight speeds between the speed for minimum required power setting and the stall speed (or minimum control speed) is termed the region of reversed command. In the region of reversed command, a decrease in airspeed must be accompanied by an increased power setting in order to maintain steady flight.

An airplane performing a low airspeed, high pitch attitude power approach for a short-field landing is an example of operating in the region of reversed command. If an unacceptably high sink rate should develop, it may be possible for the pilot to reduce or stop the descent by applying power. But without further use of power, the airplane would probably stall or be incapable of flaring for the landing. Merely lowering the nose of the airplane to regain flying speed in this situation, without the use of power, would result in a rapid sink rate and corresponding loss of altitude.

If during a soft-field takeoff and climb, for example, the pilot attempts to climb out of ground effect without first attaining normal climb pitch attitude and airspeed, the airplane may inadvertently enter the region of reversed command at a dangerously low altitude. Even with full power, the airplane may be incapable of climbing or even maintaining altitude. The pilot's only recourse in this situation is to lower the pitch attitude in order to increase airspeed, which will inevitably result in a loss of altitude. Airplane pilots must give particular attention to precise control of airspeed when operating in the low flight speeds of the region of reversed command.

11. The primary purpose of the passage is to
    a. instruct pilots on proper airspeed.
    b. discuss the interrelationships of airspeed, power, and pitch attitude.
    c. explain reversed command.
    d. discuss the physics of flight at low airspeeds.
    e. persuade the reader to fly faster aircraft.

12. In the fifth paragraph, *inadvertently* most nearly means
    a. unintentionally
    b. indirectly
    c. sequentially
    d. primarily
    e. eventually

13. In which region of command does most flight occur?
    a. inverse command
    b. normal command
    c. direct command
    d. reverse command
    e. decreased command

14. With which one of the following statements about flight would the author most likely agree?
    a. As speed increases, the power required to descend decreases.
    b. As speed increases, the power required to descend remains constant.
    c. As speed decreases, the power required to climb remains constant.
    d. As speed increases, the power required to maintain altitude increases.
    e. As speed decreases, the power required to maintain altitude increases.

15. Which of the following would be the best title for this passage?
    a. How to Avoid Reversed Command
    b. Normal Command Flight
    c. Learning to Fly
    d. Power Requirements During Flight
    e. Reversed Command and the Modern Pilot

Passage 4
In many cases, the landing distance of an aircraft will define the runway requirements for flight operations. The minimum landing distance is obtained by landing at some minimum safe speed, which allows sufficient margin above stall and provides satisfactory control and capability for a go-around. Generally, the landing speed is some fixed percentage of the stall speed or minimum control speed for the aircraft in the landing configuration. As such, the landing will be accomplished at some particular value of lift coefficient and AOA. The exact values will depend on the aircraft characteristics but, once defined, the values are independent of weight, altitude, and wind.

To obtain minimum landing distance at the specified landing speed, the forces that act on the aircraft must provide maximum deceleration during the landing roll. The forces acting on the aircraft during the landing roll may require various procedures to maintain landing deceleration at the peak value.

A distinction should be made between the procedures for minimum landing distance and an ordinary landing roll with considerable excess runway available. Minimum landing distance will be obtained by creating a continuous peak deceleration of the aircraft; that is, extensive use of the brakes for maximum deceleration. On the other hand, an ordinary landing roll with considerable excess runway may allow extensive use of aerodynamic drag to minimize wear and tear on the tires and brakes. If aerodynamic drag is sufficient to cause deceleration, it can be used in deference to the brakes in the early stages of the landing roll; i.e., brakes and tires suffer from continuous hard use, but aircraft aerodynamic drag is free and does not wear out with use.

The use of aerodynamic drag is applicable only for deceleration to 60 or 70 percent of the touchdown speed. At speeds less than 60 to 70 percent of the touchdown speed, aerodynamic drag is so slight as to be of little use, and braking must be utilized to produce continued deceleration. Since the objective during the landing roll is to decelerate, the powerplant thrust should be the smallest possible positive value (or largest possible negative value in the case of thrust reversers). In addition to the important factors of proper procedures, many other variables affect the landing performance. Any item that alters the landing speed or deceleration rate during the landing roll will affect the landing distance.

The effect of gross weight on landing distance is one of the principal items determining the landing distance. One effect of an increased gross weight is that a greater speed will be required to support the aircraft at the landing AOA and lift coefficient. For an example of the effect of a change in gross weight, a 21 percent increase in landing weight will require a ten percent increase in landing speed to support the greater weight.

When minimum landing distances are considered, braking friction forces predominate during the landing roll and, for the majority of aircraft configurations, braking friction is the main source of deceleration.

The minimum landing distance will vary in direct proportion to the gross weight. For example, a ten percent increase in gross weight at landing would cause a:

- Five percent increase in landing velocity
- Ten percent increase in landing distance

A contingency of this is the relationship between weight and braking friction force.

The effect of wind on landing distance is large and deserves proper consideration when predicting landing distance. Since the aircraft will land at a particular airspeed independent of the wind, the principal effect of wind on landing distance is the change in the groundspeed at which the aircraft touches down. The effect of wind on deceleration during the landing is identical to the effect on acceleration during the takeoff.

The effect of pressure altitude and ambient temperature is to define density altitude and its effect on landing performance. An increase in density altitude increases the landing speed but does not alter the net retarding force. Thus, the aircraft at altitude lands at the same IAS as at sea level but, because of the reduced density, the TAS is greater. Since the aircraft lands at altitude with the same weight and dynamic pressure, the drag and braking friction throughout the landing roll have the same values as at sea level. As long as the condition is within the capability of the brakes, the net retarding force is unchanged, and the deceleration is the same as with the landing at sea level. Since an increase in altitude does not alter deceleration, the effect of density altitude on landing distance is due to the greater TAS.

16. The main purpose of the passage is to
    a. give some examples of near accidents during landing.
    b. improve landing skills.
    c. explain the effects of varying pressure altitudes.
    d. advocate safer protocols for landing.
    e. describe the factors that influence landing distance.

17. Why will a pilot rely on aerodynamic drag when making a normal landing?
    a. To increase the rate of deceleration
    b. To avoid wearing down the brakes and tires
    c. To avoid unnecessary turbulence
    d. To mitigate a large gross weight
    e. To simplify landing procedures

18. In the fourth paragraph, *principal* most nearly means
    a. most important
    b. easiest
    c. moral
    d. first
    e. value

19. Why must a heavier plane land at a higher speed?
    a. To diminish fuel supplies and thereby decrease gross weight
    b. To encourage a stall just before landing
    c. To avoid hitting the runway with too much force
    d. To improve handling on the runway
    e. To allow for the longest landing distance

20. With which one of the following claims about landing performance would the author most likely agree?
    a. Many landings occur without any use of the brakes.
    b. Ambient temperature has no effect on minimum landing distance.
    c. Gross weight and minimum landing distance are positively correlated.
    d. Runway length is less important than gross weight in determining the appropriate airspeed during landing.
    e. Minimum landing distance is generally consistent for aircraft of the same size.

Passage 5

For over 25 years, the importance of good pilot judgment, or aeronautical decision-making (ADM), has been recognized as critical to the safe operation of aircraft, as well as accident avoidance. The airline industry, motivated by the need to reduce accidents caused by human factors, developed the first training programs based on improving ADM. Crew resource management (CRM) training for flight crews is focused on the effective use of all available resources: human resources, hardware, and information supporting ADM to facilitate crew cooperation and improve decision-making. The goal of all flight crews is good ADM and the use of CRM is one way to make good decisions.

Research in this area prompted the Federal Aviation Administration (FAA) to produce training directed at improving the decision-making of pilots and led to current FAA regulations that require that decision-making be taught as part of the pilot training curriculum. ADM research, development, and testing culminated in 1987 with the publication of six manuals oriented to the decision-making needs of variously rated pilots.

These manuals provided multifaceted materials designed to reduce the number of decision related accidents. The effectiveness of these materials was validated in independent studies where student pilots received such training in conjunction with the standard flying curriculum. When tested, the pilots who had received ADM training made fewer inflight errors than those who had not received ADM training. The differences were statistically significant and ranged from about 10 to 50 percent fewer judgment errors. In the operational environment, an operator flying about 400,000 hours annually demonstrated a 54 percent reduction in accident rate after using these materials for recurrency training.

Contrary to popular opinion, good judgment can be taught. Tradition held that good judgment was a natural by-product of experience, but as pilots continued to log accident-free flight hours, a corresponding increase of good judgment was assumed. Building upon the foundation of conventional decision-making, ADM enhances the process to decrease the probability of human error and increase the probability of a safe flight. ADM provides a structured, systematic approach to analyzing changes that occur during a flight and how these changes might affect a flight's safe outcome. The ADM process addresses all aspects of decision-making in the flight deck and identifies the steps involved in good decision-making.

Steps for good decision-making are:

1. Identifying personal attitudes hazardous to safe flight
2. Learning behavior modification techniques
3. Learning how to recognize and cope with stress
4. Developing risk assessment skills
5. Using all resources
6. Evaluating the effectiveness of one's ADM skills

Risk management is an important component of ADM. When a pilot follows good decision-making practices, the inherent risk in a flight is reduced or even eliminated. The ability to make good decisions is based upon direct or indirect experience and education.

Consider automotive seat belt use. In just two decades, seat belt use has become the norm, placing those who do not wear seat belts outside the norm, but this group may learn to wear a seat belt by either direct or indirect experience.

For example, a driver learns through direct experience about the value of wearing a seat belt when he or she is involved in a car accident that leads to a personal injury. An indirect learning experience occurs when a loved one is injured during a car accident because he or she failed to wear a seat belt.

While poor decision-making in everyday life does not always lead to tragedy, the margin for error in aviation is thin. Since ADM enhances management of an aeronautical environment, all pilots should become familiar with and employ ADM.

21. The primary purpose of the passage is to
   a. list the steps in good decision-making.
   b. improve the decision-making abilities of the reader.
   c. outline the relationship between aeronautical decision-making and crew resource management.
   d. discuss aeronautical decision-making.
   e. inspire the reader to make better decisions.

22. According to the passage, how is aviation safety distinguished from other forms of safety?
   a. Aviation safety is much simpler than most other areas of safety.
   b. Aviation safety is no different than most other areas of safety.
   c. Aviation safety is only important to a small percentage of the population.
   d. There is a smaller margin for error in aviation.
   e. Aviation safety can be systematized.

23. In the second paragraph, *conjunction* most nearly means
   a. combination
   b. linking word
   c. opposition
   d. collection
   e. organization

24. With which one of the following claims about aeronautical decision-making would the author most likely agree?
   a. The body of knowledge about ADM is increasing, and this will have a positive effect on flight safety.
   b. Aeronautical decision-making is the responsibility of the pilot alone.
   c. Eventually, researchers will establish a perfect set of decision-making tools for pilots.
   d. Aeronautical decision-making is the only tool required for flight safety.
   e. Aeronautical decision-making has no applications in areas other than flight.

25. When a pilot reads the account of a recent aviation accident, this is an opportunity for a(n)
   a. recertification.
   b. direct learning experience.
   c. implicit learning experience.
   d. reorientation of learning.
   e. indirect learning experience.

Situational Judgment

Situation 1:
You are approached by a senior officer, who requests a private meeting with you. He asks you for your candid opinion on your immediate supervisor, who is a subordinate to the senior officer. You have a generally favorable opinion of the supervisor, but you do have a few complaints about her performance. Specifically, you feel that she does a poor job of running staff meetings.
Possible actions:

a. Decline to meet with the senior officer.

b. Meet with the senior officer, and focus on the ways your supervisor could improve staff meetings.

c. Give your candid opinion of your supervisor, including your criticisms, but emphasizing your overall positive opinion.

d. Write a letter to the senior officer, explaining your opinions about your supervisor.

e. Give the senior officer a glowing report of the supervisor, without mentioning your complaints.

1. Select the MOST EFFECTIVE action in response to the situation.
2. Select the LEAST EFFECTIVE action in response to the situation.

Situation 2:
While performing administrative work with another officer, you notice that he is manipulating the numbers on some reports. Specifically, the officer is inflating the amounts of time spent on certain training exercises. These reports are sent on to senior officers, who use them to assess the readiness of the soldiers for more sophisticated and complicated missions. If men and women are unprepared for these more difficult missions, there is a greater risk of accident or injury, to themselves or others.
Possible actions:

a. Immediately report this infraction to your superior.

b. Wait until you can acquire clear proof of the data manipulation.

c. Say nothing at present, but keep an eye on this other officer.

d. Write an anonymous letter to the other officer, encouraging him to stop manipulating the data.

e. Go back over the other officer's work, correcting the data as necessary.

3. Select the MOST EFFECTIVE action in response to the situation.
4. Select the LEAST EFFECTIVE action in response to the situation.

Situation 3:
You and two of your fellow officers have been selected to interview candidates for a promotion. One of the three candidates (Candidate A) is known to be an old family friend of your commanding officer. During the interviews, Candidate A is, in your opinion, the most impressive. However, Candidate B also seems like an excellent choice for the position. Candidate C performs poorly. One of the other judges votes for Candidate A, and the other votes for Candidate B.
Possible actions:

a. Vote for Candidate B to avoid the appearance of favoritism.
b. Vote for Candidate C so that the decision will be passed on to another round of deliberation.
c. Abstain from voting.
d. Request a new set of candidates.
e. Vote for Candidate A because you feel he is the most qualified.

5. Select the MOST EFFECTIVE action in response to the situation.
6. Select the LEAST EFFECTIVE action in response to the situation.

Situation 4:
You attend a meeting with two other officers. During the meeting, these officers get into a heated conflict over a possible change in policy. You have heard that they have a personal antipathy, though you are not aware of its origins. They ask you to settle their dispute.
Possible actions:

a. Ignore the request, and instead lecture the two officers on the need for cooperation and goodwill in the military.
b. Take the side of the officer who you believe will be the most help to you in the future.
c. Refuse to take a side, citing the obvious personal differences between the two officers.
d. Select the option that you think is best, leaving aside everything you know about their personal conflict.
e. Take the side of the officer you like better.

7. Select the MOST EFFECTIVE action in response to the situation.
8. Select the LEAST EFFECTIVE action in response to the situation.

Situation 5:
You are asked to collaborate on an important project with an officer from a different unit. This officer is very close to retirement, and appears to have lost interest in his work. Consequently, he puts very little effort into the work you two are supposed to be sharing.
Possible actions:

a. Request a different partner for the project.
b. Inform your senior officer of the situation.
c. Do your best on your work, but accept that the project probably will fail because of the poor attitude of your collaborator.
d. Do your work and the work that was supposed to be done by your partner, since the most important thing is the successful completion of the project.
e. Express your frustrations directly to your collaborator, emphasizing that the success of this project is very important to you, and agree upon a fair division of labor.

9. Select the MOST EFFECTIVE action in response to the situation.
10. Select the LEAST EFFECTIVE action in response to the situation.

Situation 6:
While working in a field office, you observe that one of your colleagues is being severely overworked. Despite doing an excellent job and working more than the required number of hours, she is unable to keep up with the amount of paperwork being sent to her by other offices. Her commanding officers do not seem to be aware that she is being asked to do an unfair amount of work.
Possible actions:

a. Assign one of your subordinates to assist your overworked colleague.
b. Request a meeting with your overworked colleague's commanding officer, and describe the problem to him.
c. Ignore the problem, since it does not relate to your work.
d. Express your sympathy with your overworked colleague.
e. Help your overworked colleague whenever you finish your own work.

11. Select the MOST EFFECTIVE action in response to the situation.
12. Select the LEAST EFFECTIVE action in response to the situation.

Situation 7:
You have become suspicious that one of your junior officers is trying to undermine your work. This junior officer is very competent and very ambitious. You have even heard from other officers that this junior officer wants to take over your job. You have not yet mentioned your suspicion directly to the junior officer.
Possible actions:

a. Meet with the junior officer and explain that you expect his full cooperation and support.
b. Publicly reprimand and humiliate the junior officer.
c. Ignore the situation, in the hopes that it will resolve itself.
d. Report this insubordination to your commanding officer.
e. Ask one of your fellow officers to talk with the junior officer and try to improve the situation.

13. Select the MOST EFFECTIVE action in response to the situation.
14. Select the LEAST EFFECTIVE action in response to the situation.

Situation 8:
You are extremely busy with paperwork, but a senior officer asks you to complete a set of special reports in addition to your normal work. You do not think it will be possible for you to complete this extra assignment without sacrificing the quality of your work. However, you would like to impress your senior officer by fulfilling his request.
Possible actions:

a. Tell the senior officer that you are too busy to complete extra projects.
b. Accept the extra work and resolve to complete it quickly no matter what.
c. Ask the senior officer if you can have a few days to decide whether you will complete the extra work.
d. Accept the extra work, but have a junior staff member complete it for you.
e. Decline the extra work, but offer to pass it along to another qualified member of your team.

15. Select the MOST EFFECTIVE action in response to the situation.
16. Select the LEAST EFFECTIVE action in response to the situation.

Situation 9:

Over the past few months, you have noticed that office supplies are being used at a much greater pace than is usual. There is no clear reason for this, but you have begun to suspect that one of your fellow officers is taking office supplies home with her at night. You have no specific evidence to support your claims, but the office supplies seem to be disappearing after shifts in which this officer is alone with access to the supply closet.

Possible actions:

a. Ignore the situation, since you know that you are not personally responsible for the thefts.
b. Tell a senior officer about your suspicions.
c. Confront the officer with your suspicions and ask for an explanation.
d. Set up a hidden surveillance camera so that you can catch the officer stealing supplies.
e. Ask some of your fellow officers if they have noticed any suspicious behavior.

17. Select the MOST EFFECTIVE action in response to the situation.
18. Select the LEAST EFFECTIVE action in response to the situation.

Situation 10:

You have been working in the same unit for the past two years. Although you have been successful, you are beginning to get burnt out, and you are thinking about requesting a transfer. You contact some friends and colleagues in other units, trying to determine whether you would like it there. After a few weeks, you realize that rumors about your interest in a transfer have begun to spread throughout your unit.

Possible actions:

a. In a letter to your senior officer, acknowledge the truth of the rumors and request a transfer.
b. Deny the rumors publicly, but continue to pursue a transfer.
c. Ignore the rumors, but refocus on your work with your current unit.
d. Acknowledge the rumors and refocus on your work with your current unit.
e. Request a leave of absence so that you can decide how to handle the situation.

19. Select the MOST EFFECTIVE action in response to the situation.
20 Select the LEAST EFFECTIVE action in response to the situation.

Situation 11:
You have been asked to attend a meeting between the senior officers in your unit and a group of local community leaders. The meeting has been called because of rumors that your military base will be shut down due to budgetary constraints. The community leaders are willing to lobby on behalf of the base, so long as they receive assurances that the military leadership will cooperate with them on some local initiatives. Specifically, the community leaders would like the soldiers to help coordinate disaster relief efforts when necessary. At one point, a community leader turns to you and asks for your opinion on the subject.
Possible actions:

a. Answer the question, but mention only the potential positive consequences of the proposal.
b. Try to answer the question as honestly and completely as possible, but remind the community leader that you are only a junior officer.
c. Answer the question thoroughly, even though you have little direct knowledge of the situation.
d. Remind the community leader that you are a junior officer and cannot give your opinion.
e. Decline to answer the question, and instead refer it to one of the senior officers.

21. Select the MOST EFFECTIVE action in response to the situation.
22. Select the LEAST EFFECTIVE action in response to the situation.

Situation 12:
Another officer in your section is granted a week-long leave to visit his sick grandmother. However, during this period you discover through social media that the officer is actually on a beach vacation with his girlfriend, hundreds of miles from the hospital where his grandmother was supposedly being treated.
Possible actions:

a. Tell the officer that he has been dishonest and unethical, and that you will have to report any future episodes of this nature.
b. Ignore the situation, since it does not really involve you.
c. Tell the officer that you will agree to keep his misbehavior a secret if he will do your weekly reports.
d. Immediately notify your commanding officer in person.
e. Anonymously forward the evidence of the officer's misbehavior to your commanding officer.

23. Select the MOST EFFECTIVE action in response to the situation.
24. Select the LEAST EFFECTIVE action in response to the situation.

Situation 13:

One of the soldiers in your unit has displayed a marked decline in her performance over the past month. She appears frustrated and burnt out with her normal duties. She is one of the more popular members of the unit, so her negative attitude has a bad influence on her fellow soldiers.

Possible actions:

a. Convene a meeting of the entire unit, and use this as a chance to single out the soldier for her poor performance.

b. Meet with the soldier and offer whatever help you can give to improve her performance, while emphasizing the effect that her behavior has on the other soldiers.

c. Reassign the soldier to a different unit so that her bad attitude doesn't continue to affect the other soldiers.

d. Do nothing, in the hopes that the situation will improve without your influence.

e. Inform your senior officer about the situation.

25. Select the MOST EFFECTIVE action in response to the situation.
26. Select the LEAST EFFECTIVE action in response to the situation.

Situation 14:

You are about to transfer to a new unit, where you will have duties in areas where you have little experience. A week before you are due to make the transfer, you receive an email from the senior officer in charge of your new unit. She reminds you that you will be entering her unit at a very important time for them, because they will be leading a set of training exercises for highly-skilled soldiers. She wants to make sure that you are ready to contribute immediately to the success of the unit, and that you will not need a great deal of assistance to complete your work.

Possible actions:

a. Do not respond to the email, and assume that you will be able to move into your new role seamlessly.

b. Thank the senior officer for her message, and do some internet research on your new duties.

c. Request that the transfer be canceled, and remain with your original unit.

d. Email the senior officer back, requesting a personal meeting where you can get more information about how to make a good transition to your new role.

e. Ask one of the other officers in your new unit if he will quietly bring you up to speed when you arrive.

27. Select the MOST EFFECTIVE action in response to the situation.
28. Select the LEAST EFFECTIVE action in response to the situation.

Situation 15:
One of your responsibilities is to keep a set of officers briefed on some confidential activities that are taking place at your base. One day, you accidentally send an email containing some information about these confidential activities to an officer who has not received the security clearance.
Possible actions:

a. Notify your supervisor of your mistake and let him resolve the situation.
b. Immediately send a second email to the improper recipient, requesting that he or she destroy the email. Inform your supervisors of your mistake.
c. Send a second email to the improper recipient, claiming that your email account has been hacked and that he should disregard any earlier messages.
d. Wait to see if there will be any negative consequences of your mistake.
e. Ask your supervisor if the improper recipient could be given the security clearance retroactively.

29. Select the MOST EFFECTIVE action in response to the situation.
30. Select the LEAST EFFECTIVE action in response to the situation.

Situation 16:
After completing your work at the base, you return to your living quarters for the evening. An hour later, you realize that you failed to affix your signature to a set of papers that are to be forwarded on to a different unit for completion. Without your signature, the papers cannot be sent. The content of the papers is not particularly urgent, but the delay will require the other unit officer to stay later than normal at his post.
Possible actions:

a. Wait until your next shift to sign the papers, since they are not considered urgent.
b. Call the base and ask a junior officer to forge your signature so the paperwork can be sent.
c. Call the officer at the other base and explain that the papers will be arriving a little later than expected.
d. Arrive for your next shift a little early and sign the papers first.
e. Return to the base and sign the necessary papers.

31. Select the MOST EFFECTIVE action in response to the situation.
32. Select the LEAST EFFECTIVE action in response to the situation.

Situation 17:

Several of the soldiers in your unit have yet to complete a basic training module at a nearby base. They have asked to participate in the next training session, but instead a group of soldiers from another unit have been selected. The selected soldiers have not been waiting nearly as long as your soldiers to complete this training module. You suspect that the director of the training session dislikes you personally, though you have no specific evidence of this.

Possible actions:

a. Request a meeting with the training director, and ask why your soldiers have been passed over, emphasizing the importance of this module for their development.

b. Do nothing, in the hopes that the situation will improve on its own.

c. Ask the officer in charge of the other unit if your soldiers can attend the training session instead of his.

d. Write a critical letter to the training director and your senior officer, outlining what you perceive as the injustice of the situation.

e. Ask the training director if your soldiers can attend the next training session along with the selected soldiers.

33. Select the MOST EFFECTIVE action in response to the situation.
34. Select the LEAST EFFECTIVE action in response to the situation.

Situation 18:

You have been assigned to draft an important report along with another officer in your unit. It is expected that the report will take approximately one month to complete. However, after about a week of work, the other officer falls ill and is required to go on leave. He is only supposed to be gone for about ten days, but at the end of this period he has not returned and there is no definitive word on when he will. You need the expertise of this officer in order to finish the report.

Possible actions:

a. Wait until the other officer returns from leave, even if it means delaying the report.

b. Ask for an extension, without making excuses for your failure to complete the report on time.

c. Work extra hours to complete the report as best as you can.

d. Tell your commanding officer the situation, and request assistance in completing the report.

e. Order a junior officer to assist you in the completion of the report.

35. Select the MOST EFFECTIVE action in response to the situation.
36. Select the LEAST EFFECTIVE action in response to the situation.

Situation 19:
Two months ago, you joined a new unit. The leader of this unit was very welcoming to you, and made sure to give you as much assistance as you needed in learning your new duties and responsibilities. However, you are now feeling more comfortable in your role and would like more independence in your work.
Possible actions:

a. Request a meeting with the unit leader, thank him for his assistance, and indicate that you would like to work on your own a bit more.
b. Request that another officer be transferred to your unit, so that the unit leader will divert his attention to training this new arrival.
c. Without saying anything directly, try to avoid the unit leader as much as possible.
d. Tell the unit leader's commanding officer that you need the unit leader to give you more space.
e. Keep extensive records of your work, so that you can demonstrate your competence to the unit leader.

37. Select the MOST EFFECTIVE action in response to the situation.
38. Select the LEAST EFFECTIVE action in response to the situation.

Situation 20:
One of the other officers in your division is going to make an important presentation at the end of the week. He is nervous, but has done good work in the past and has spent a great deal of time preparing the report. He asks you to look over his work in advance. You notice a few things that need to be changed, but the other officer disagrees with your corrections. You are certain that you are right and that the other officer will be sorry he did not listen to you.
Possible actions:

a. Let your coworker go ahead with the uncorrected presentation, but make yourself look better by mentioning the errors to a senior officer before the presentation.
b. Allow your coworker to go ahead with the presentation as is, without trying to convince him to make the corrections.
c. Discuss the matter with your senior officer, and ask him or her to mandate the corrections.
d. Contrive an excuse to be absent from the presentation.
e. Make every effort to convince your colleague to make the necessary corrections.

39. Select the MOST EFFECTIVE action in response to the situation.
40. Select the LEAST EFFECTIVE action in response to the situation.

Situation 21:
You have been working with a new unit for the past six weeks. During that time, you have noticed some inefficiencies in the unit's operations, and you have developed a set of proposals for eliminating them. The majority of your coworkers agree with your proposals, but the senior officer in charge of the unit does not. The senior officer believes that implementing your proposals would be too risky and would undermine the stability of the unit.
Possible actions:

a. Confront your senior officer, using the support of your coworkers as a reason to implement your proposals.
b. Create a detailed and comprehensive report outlining the potential benefits of your proposals. Deliver the report and then obey the senior officer's final decision.
c. Implement your proposals anyway, on the assumption that your senior officer will change his mind once he sees their success.
d. Accept the senior officer's decision and try to succeed within the agreed-upon structure.
e. Accept the senior officer's decision, but keep a running list of the ways your proposals could have improved performance, had they been implemented.

41. Select the MOST EFFECTIVE action in response to the situation.
42. Select the LEAST EFFECTIVE action in response to the situation.

Situation 22:
Six months ago, you were assigned a new assistant. Although you have been able to work successfully together, you have developed a personal dislike for this person. In your opinion, he is arrogant and too critical of the other officers. During a meeting with your senior officer, she mentions that she is considering transferring your assistant to another unit. This would represent a step up for him, and would make it possible for him to achieve even more promotions in a relatively short time.
Possible actions:

a. Recommend that your assistant receive the promotion, if only to get him away from you.
b. Strongly discourage the senior officer from choosing your assistant, so that he will not get a professional reward.
c. Write an anonymous letter to the senior officer, outlining your complaints about your assistant.
d. Strongly discourage the senior officer from choosing your assistant, with an emphasis on his personality flaws.
e. Avoid interfering in the senior officer's decision, but make sure that she is aware of your opinion of your assistant.

43. Select the MOST EFFECTIVE action in response to the situation.
44. Select the LEAST EFFECTIVE action in response to the situation.

Situation 23:
Your base uses a computer program to determine the logistics related to supply deliveries. One day, while you are coordinating the arrival and unloading of several concurrent deliveries, the computer system crashes. The computer technician tells you that it could be an hour before the system is up and running again. You can see that there is a long line of trucks waiting to deliver their goods, and that the drivers are becoming impatient.
Possible actions:

a. Ask one of your assistant to inform the drivers of the situation and the likely wait time. Offer whatever accommodations you can in the interim.
b. Receive the deliveries despite the computer problems, and keep paper records so you can update the system later.
c. Call your senior officer and ask for advice.
d. Use this opportunity to take your lunch break, somewhere you are unlikely to meet any of the drivers.
e. Encourage the drivers to make any other deliveries they have and then come back later.

45. Select the MOST EFFECTIVE action in response to the situation.
46. Select the LEAST EFFECTIVE action in response to the situation.

Situation 24:
You have developed an idea that you believe will improve the performance of your unit. However, some of the soldiers in your unit disagree with this idea, and one has gone so far as to write a letter of complaint to your senior officer without notifying you. You have not yet implemented your idea.
Possible actions:

a. Meet with the letter-writer and other critics, emphasizing that going above your head will not be tolerated in the future.
b. Abandon your idea and ask for suggestions from the soldiers who were critical of it.
c. Ignore the critics in you unit, and implement your idea anyway.
d. Harshly punish the letter writer, as a warning to the other soldiers.
e. Implement your idea without acknowledging the letter or the criticism of other soldiers in the unit.

47. Select the MOST EFFECTIVE action in response to the situation.
48. Select the LEAST EFFECTIVE action in response to the situation.

Situation 25:
You received a promotion six months ago, and have been excelling in your new job. However, due to forces beyond your control, the quality of your work has decreased over the past few weeks. In part, you have been undermined by recent budget cuts. Unfortunately, your commanding officer does not fully understand the consequences of the budget cuts, and has expressed her displeasure with your recent work. She has even suggested that problems in your unit may be a result of poor management on your part.
Possible actions:

a. Ask the commanding officer if you can take a brief leave to refocus.
b. Do not make any excuses, but ask the advice of other officers who are dealing with the same budget constraints.
c. Ask your commanding officer for a list of her specific complaints.
d. Shift the blame for your unit's performance to your subordinates.
e. Remind the commanding officer of the budget constraints, and defend your management style to her.

49. Select the MOST EFFECTIVE action in response to the situation.
50. Select the LEAST EFFECTIVE action in response to the situation.

Physical Science

1. A long nail is heated at one end. After a few seconds, the other end of the nail becomes equally hot. What type of heat transfer does this represent?
   a. Advection
   b. Conduction
   c. Convection
   d. Entropy
   e. Radiation

2. The measure of energy within a system is called:
   a. temperature
   b. convection
   c. entropy
   d. thermodynamics
   e. heat

3. How do two isotopes of the same element differ?
   a. They have different numbers of protons
   b. They have different numbers of neutrons
   c. They have different numbers of electrons
   d. They have different charges
   e. They have different atomic numbers

4. Which type of nuclear process features atomic nuclei splitting apart to form smaller nuclei?
   a. Fission
   b. Fusion
   c. Decay
   d. Ionization
   e. Chain reaction

5. The process whereby a radioactive element releases energy slowly over a long period of time to lower its energy and become more stable is best described as:
   a. combustion
   b. fission
   c. fusion
   d. decay
   e. radioactivity

6. What property of light explains why a pencil in a glass of water appears to be bent?
   a. reflection
   b. refraction
   c. angle of incidence = angle of reflection
   d. constructive interference
   e. destructive interference

7. What unit describes the frequency of a wave?
   a. hertz (Hz)
   b. decibels (dB)
   c. meters (m)
   d. meters per second (m/s)
   e. meters per second squared (m/s^2)

8. Which of the following is an example of kinetic energy being converted to potential energy?
   a. A child sliding down a slide.
   b. A cyclist coasting on his way up a hill.
   c. A pilot deploying airbrakes on approach to land.
   d. A motorist swerving to avoid a deer.
   e. A pair of billiard balls colliding and rebounding off each other.

9. The boiling of water is an example of:
   a. sublimation.
   b. condensation.
   c. neutralization.
   d. chemical change
   e. physical change

10. The center of an atom is called the:
   a. nucleus
   b. nuclide
   c. neutrino
   d. electron cloud
   e. electrolyte

11. When a solid is heated and transforms directly to the gaseous phases, this process is called:
   a. sublimation
   b. fusion
   c. diffusion
   d. condensation
   e. fission

12. Which scientist was responsible for developing the format of the modern periodic table?
   a. Faraday
   b. Einstein
   c. Hess
   d. Mendeleev
   e. Oppenheimer

13. The density of a material refers to its:
   a. Mass per unit volume
   b. Mass per unit length
   c. Mass per unit surface area
   d. Volume per unit surface area
   e. Volume per unit length

14. The precision of a set of experimentally obtained data points refers to:
  a. How accurate the data points are
  b. How many errors the data points contain
  c. How close the data points are to the mean of the data
  d. How close the data points are to the predicted result
  e. How close the set of data is to a normal distribution

15. Current, or the amount of electricity that is flowing, is measured in:
  a. volts
  b. watts
  c. ohms
  d. farads
  e. amperes

16. A solar eclipse can only occur if:
  a. the earth and the sun are on the same side of the moon
  b. the earth is between the sun and the moon
  c. the moon is between the earth and the sun
  d. the sun is between the earth and the moon
  e. the moon is full

17. What property of motion explains why passengers in a turning car feel pulled toward the outside of the turn?
  a. centripetal force
  b. inertia
  c. normal force
  d. impulse
  e. torque

18. According to the Ideal Gas Law, if a certain amount of gas is being held at a constant volume, and the temperature is increased, what will happen?
  a. the mass of the gas will increase
  b. the pressure of the gas will increase
  c. the density of the gas will decrease
  d. the mass of the gas will decrease
  e. the pressure of the gas will decrease

19. What wave characteristic is related to the loudness of a sound?
  a. frequency
  b. amplitude
  c. wavelength
  d. velocity
  e. period

20. Which of the following scenarios is NOT an example of a person applying work to a book?
  a. A book is picked up from the floor and put on a shelf.
  b. A book is pushed across a table top.
  c. A backpack holding a book is carried across the room.
  d. A book is held and then released so that it falls to the ground.
  e. A book is thrown vertically into the air.

Table Reading

For each question, select the number that appears in the table at the given coordinates. Recall that the first number in the ordered pair gives the column number, and the second gives the row number. For instance, the ordered pair (2, -1) refers to the number in column 2, row -1, which is 76 in the first table.

Use the table on the right to answer questions 1-5.

	-3	-2	-1	0	1	2	3
3	29	11	43	26	35	31	39
2	67	82	86	99	35	52	63
1	34	37	91	73	92	49	32
0	45	55	95	52	92	94	22
-1	51	96	39	96	79	76	34
-2	23	54	55	60	68	43	67
-3	49	59	21	68	97	82	93

1. (2, 2)
   a. 35
   b. 39
   c. 59
   d. 45
   e. 52

2. (3, -2)
   a. 96
   b. 34
   c. 49
   d. 67
   e. 23

3. (1, 1)
   a. 92
   b. 99
   c. 76
   d. 91
   e. 32

4. (0, -2)
   a. 60
   b. 55
   c. 67
   d. 49
   e. 68

5. (3, -3)
   a. 59
   b. 92
   c. 39
   d. 51
   e. 93

Use the table on the right to answer questions 6-10.

6. (-3, 1)
   a. 55
   b. 68
   c. 23
   d. 70
   e. 52

7. (3, 2)
   a. 58
   b. 67
   c. 91
   d. 11
   e. 52

8. (-2, -3)
   a. 31
   b. 39
   c. 63
   d. 79
   e. 43

9. (2, -1)
   a. 53
   b. 96
   c. 32
   d. 31
   e. 35

10. (-1, 0)
   a. 55
   b. 63
   c. 56
   d. 31
   e. 95

	-3	-2	-1	0	1	2	3
3	68	11	16	37	26	78	77
2	25	97	79	97	79	19	58
1	70	38	15	22	37	63	90
0	29	99	56	80	96	39	44
-1	29	13	18	53	44	53	47
-2	90	13	86	31	12	11	26
-3	18	63	61	78	64	86	61

Use the table on the right to answer questions 11-15.

	-3	-2	-1	0	1	2	3
3	22	14	57	62	20	16	87
2	63	90	60	77	25	86	42
1	69	78	35	54	57	32	29
0	81	97	37	72	84	39	48
-1	59	95	74	61	52	40	44
-2	99	17	93	89	25	75	19
-3	10	16	59	26	84	10	39

11. (1, 2)
    a. 32
    b. 73
    c. 67
    d. 55
    e. 25

12. (0, 2)
    a. 22
    b. 67
    c. 49
    d. 93
    e. 77

13. (-2, -2)
    a. 17
    b. 21
    c. 96
    d. 82
    e. 23

14. (0, 0)
    a. 68
    b. 96
    c. 73
    d. 72
    e. 34

15. (1, 3)
    a. 20
    b. 22
    c. 54
    d. 45
    e. 49

Use the table on the right to answer questions 16-20.

	-3	-2	-1	0	1	2	3
3	38	70	36	61	88	20	86
2	42	74	20	79	38	55	58
1	64	18	31	42	74	76	54
0	39	44	66	66	84	95	40
-1	23	73	98	14	58	86	80
-2	41	84	91	49	60	97	52
-3	85	11	51	50	14	21	99

16. (1, 1)
    a. 99
    b. 74
    c. 63
    d. 49
    e. 52

17. (0, 0)
    a. 68
    b. 45
    c. 66
    d. 35
    e. 93

18. (2, 2)
    a. 55
    b. 73
    c. 35
    d. 97
    e. 21

19. (0, 1)
    a. 86
    b. 54
    c. 42
    d. 35
    e. 94

20. (1, -1)
    a. 76
    b. 58
    c. 86
    d. 54
    e. 82

Use the table on the right to answer questions 21-25.

21. (3, -3)
    a. 94
    b. 68
    c. 25
    d. 73
    e. 79

22. (-3, -3)
    a. 26
    b. 42
    c. 52
    d. 68
    e. 31

23. (2, -2)
    a. 31
    b. 68
    c. 35
    d. 95
    e. 39

24. (-2, 0)
    a. 72
    b. 34
    c. 68
    d. 22
    e. 11

25. (-2, -1)
    a. 82
    b. 39
    c. 49
    d. 99
    e. 55

	-3	-2	-1	0	1	2	3
3	64	53	66	11	83	67	87
2	31	28	87	79	21	73	19
1	27	11	20	82	22	52	68
0	97	72	73	16	59	37	13
-1	96	49	68	96	97	42	53
-2	74	48	49	35	66	39	36
-3	42	69	71	95	60	31	25

Use the table on the right to answer questions 26-30.

	-3	-2	-1	0	1	2	3
3	89	29	33	74	87	21	61
2	42	77	18	45	16	64	97
1	22	24	28	57	67	92	78
0	49	23	14	11	92	54	24
-1	96	10	20	40	54	66	68
-2	35	79	42	95	60	34	44
-3	91	85	67	40	60	44	58

26. (0, 0)
    a. 35
    b. 76
    c. 11
    d. 86
    e. 29

27. (-1, -1)
    a. 51
    b. 96
    c. 97
    d. 60
    e. 20

28. (2, 0)
    a. 37
    b. 54
    c. 82
    d. 55
    e. 43

29. (0, -3)
    a. 96
    b. 26
    c. 40
    d. 39
    e. 54

30. (1, 0)
    a. 52
    b. 86
    c. 55
    d. 92
    e. 99

Use the table on the right to answer questions 31-35.

31. (1, -3)
    a. 68
    b. 82
    c. 60
    d. 26
    e. 34

32. (3, 0)
    a. 94
    b. 93
    c. 21
    d. 49
    e. 92

33. (-3, 0)
    a. 11
    b. 35
    c. 20
    d. 34
    e. 49

34. (3, -1)
    a. 67
    b. 43
    c. 96
    d. 29
    e. 81

35. (1, -1)
    a. 67
    b. 79
    c. 55
    d. 23
    e. 35

	-3	-2	-1	0	1	2	3
3	97	75	86	89	22	31	43
2	62	30	48	35	34	82	78
1	59	25	97	99	57	89	54
0	20	88	57	70	67	71	21
-1	84	65	50	35	67	32	81
-2	19	21	74	30	84	75	69
-3	98	19	37	47	68	14	24

Use the table on the right to answer questions 36-40.

	-3	-2	-1	0	1	2	3
3	54	54	87	36	15	61	94
2	71	79	75	70	49	92	59
1	18	43	92	70	98	66	63
0	97	57	35	58	25	67	79
-1	34	59	18	62	87	44	31
-2	95	97	65	45	88	51	10
-3	38	39	88	95	22	19	49

36. (1, -3)
    a. 43
    b. 95
    c. 22
    d. 37
    e. 91

37. (-3, 0)
    a. 97
    b. 37
    c. 43
    d. 99
    e. 35

38. (-2, -1)
    a. 59
    b. 51
    c. 92
    d. 26
    e. 23

39. (-1, 3)
    a. 96
    b. 76
    c. 87
    d. 54
    e. 34

40. (2, 2)
    a. 67
    b. 52
    c. 60
    d. 92
    e. 35

Instrument Comprehension

1. Which of the answer choices represents the orientation of the plane?

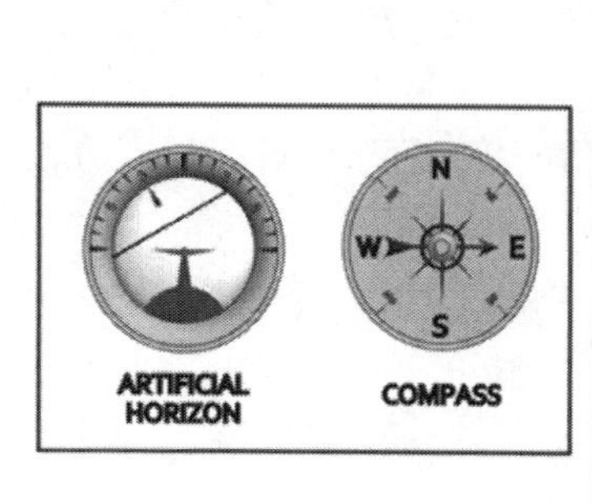

2. Which of the answer choices represents the orientation of the plane?

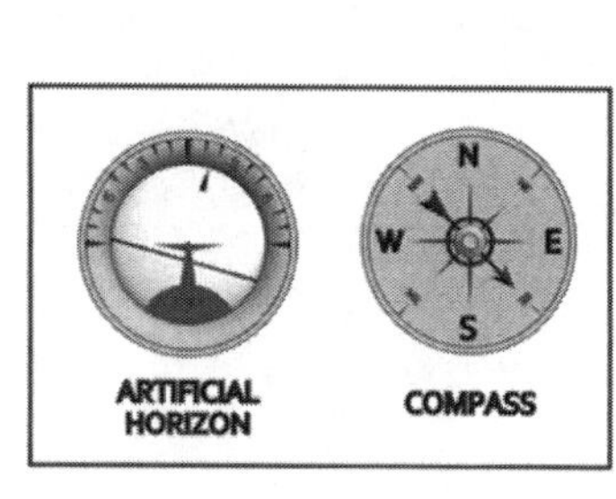

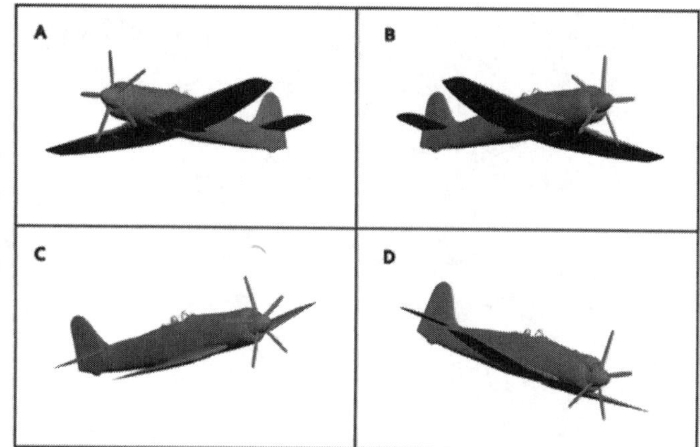

3. Which of the answer choices represents the orientation of the plane?

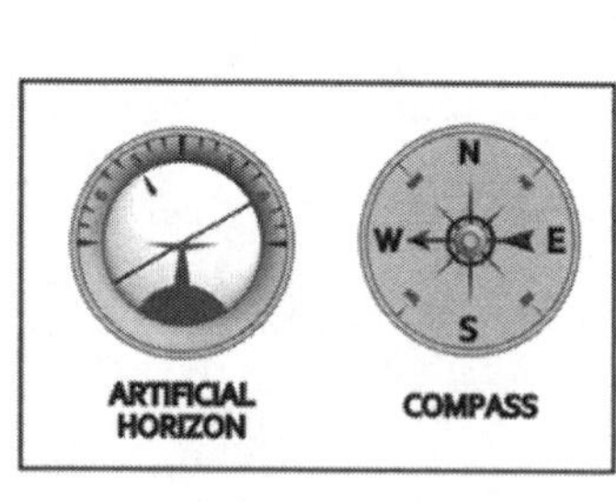

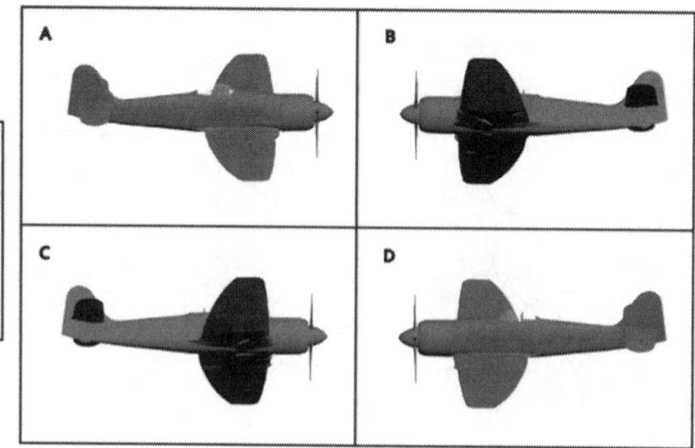

4. Which of the answer choices represents the orientation of the plane?

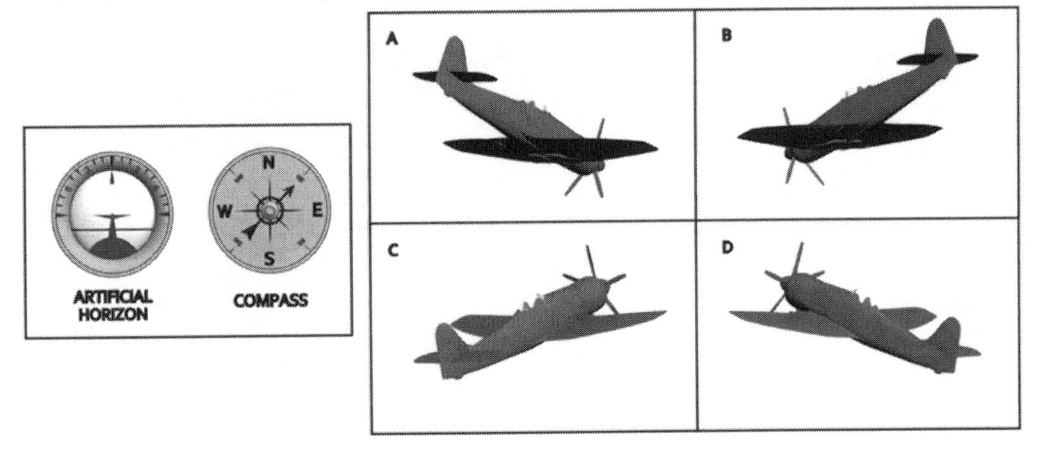

5. Which of the answer choices represents the orientation of the plane?

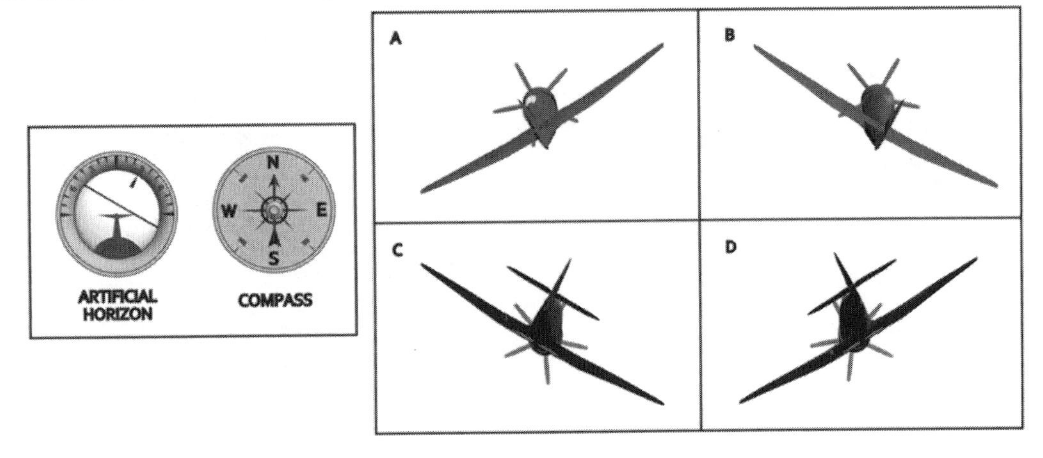

6. Which of the answer choices represents the orientation of the plane?

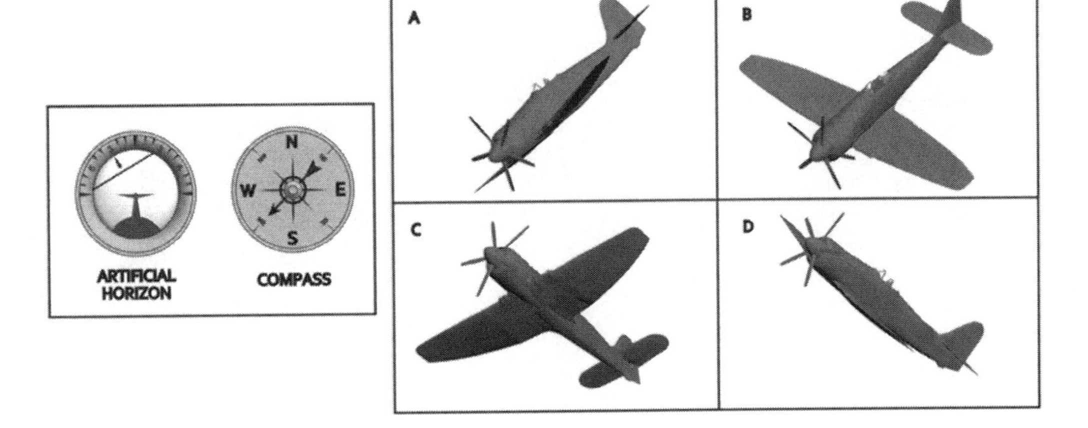

7. Which of the answer choices represents the orientation of the plane?

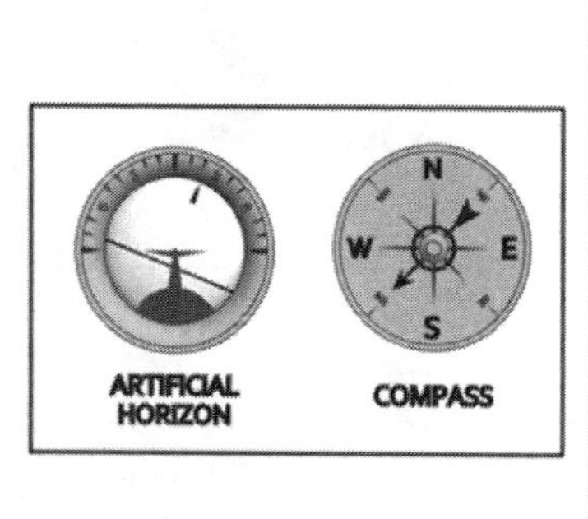

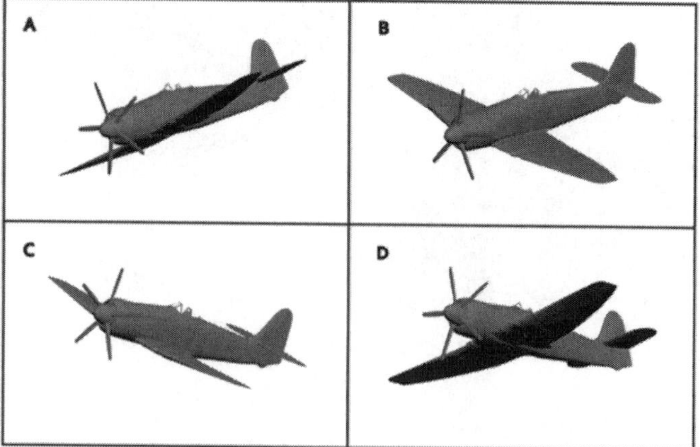

8. Which of the answer choices represents the orientation of the plane?

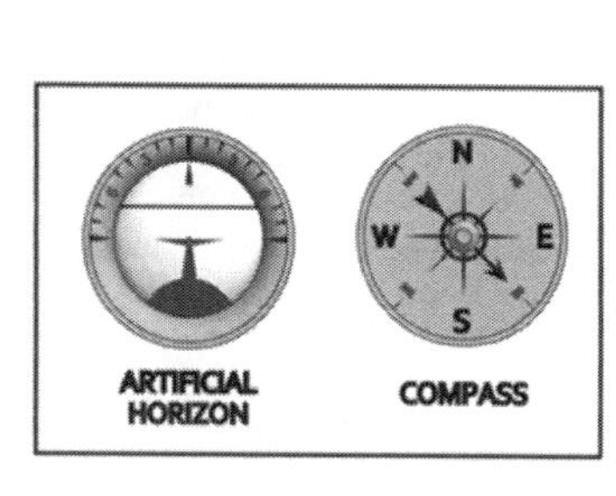

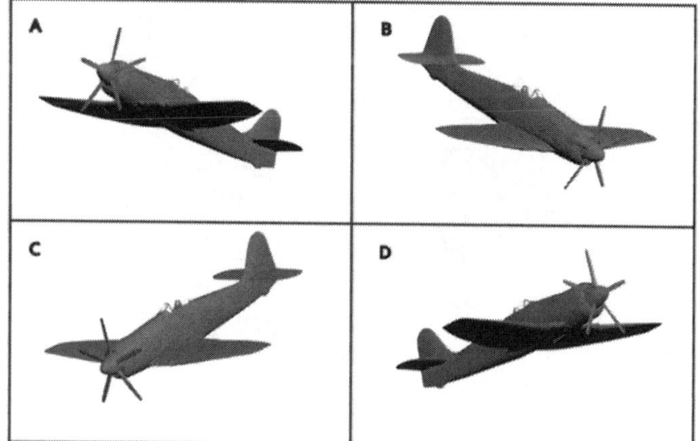

9. Which of the answer choices represents the orientation of the plane?

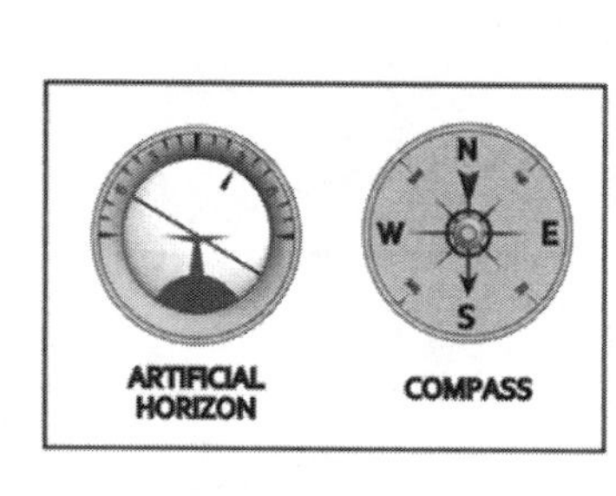

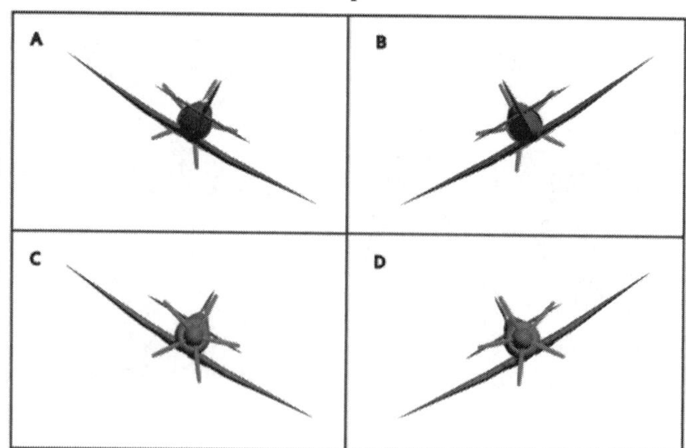

10. Which of the answer choices represents the orientation of the plane?

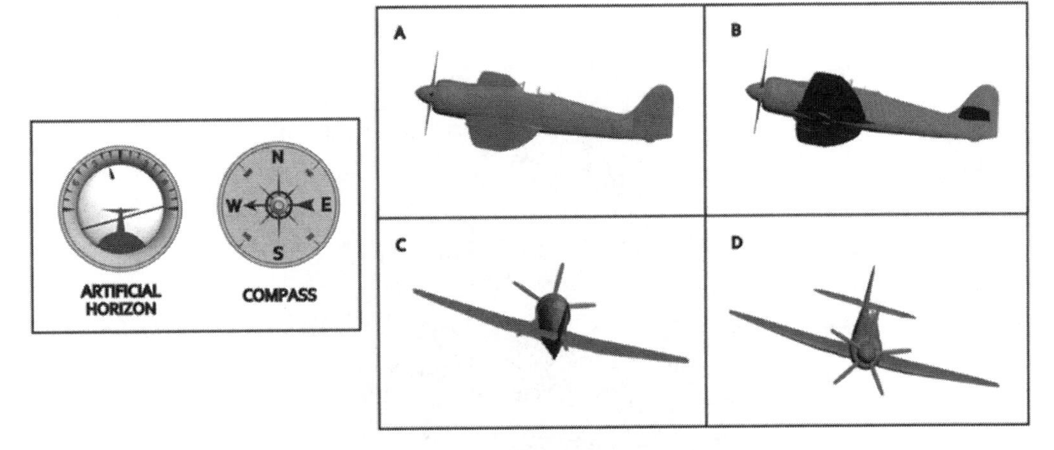

11. Which of the answer choices represents the orientation of the plane?

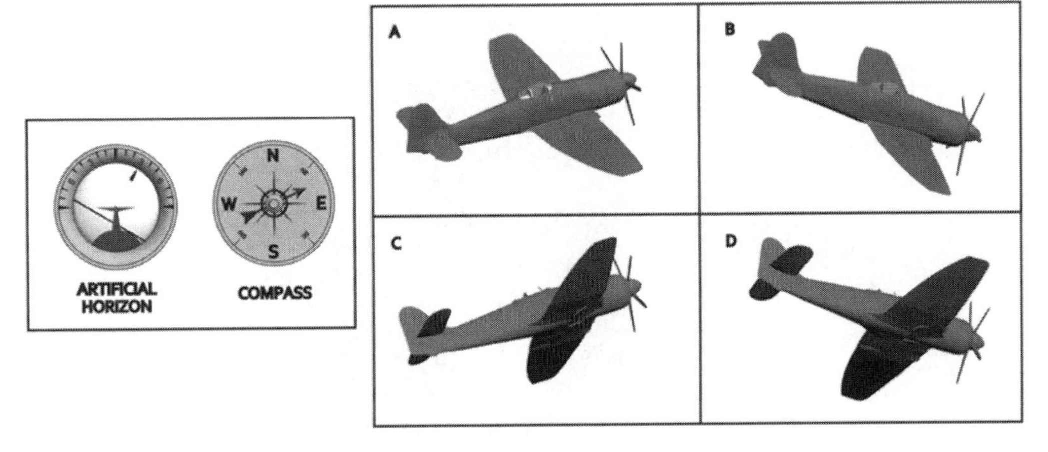

12. Which of the answer choices represents the orientation of the plane?

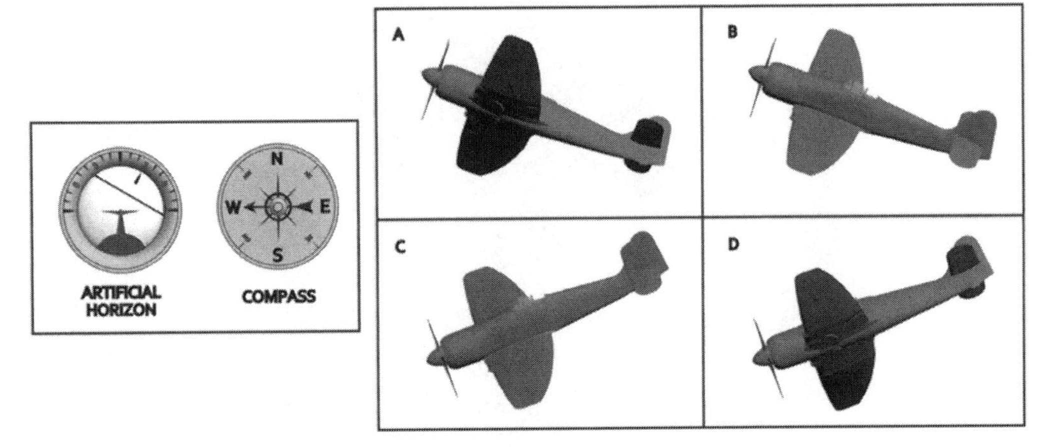

Block Counting

1. How many blocks are touching block 5 in the figure above?

2. How many blocks are touching block 1 in the figure above?

3. How many blocks are touching block 3 in the figure above?

4. How many blocks are touching block 6 in the figure above?

5. How many blocks are touching block 7 in the figure above?

6. How many blocks are touching block 9 in the figure above?

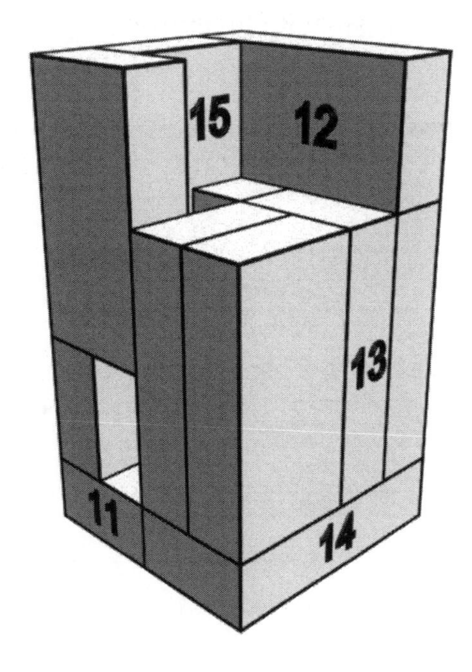

7. How many blocks are touching block 15 in the figure above?

8. How many blocks are touching block 11 in the figure above?

9. How many blocks are touching block 13 in the figure above?

10. How many blocks are touching block 16 in the figure above?

11. How many blocks are touching block 19 in the figure above?

12. How many blocks are touching block 20 in the figure above?

Aviation Information

1. From approximately how far away should a Visual Approach Slope Indicator be visible at night?
    a. Five miles
    b. Ten miles
    c. Twenty miles
    d. Thirty miles
    e. Fifty miles

2. Which of the following is NOT one of the forces a pilot must manage during flight?
    a. thrust
    b. gravity
    c. drag
    d. lift
    e. torque

3. Which control affects the angle of the main rotor blades of a helicopter?
    a. collective
    b. throttle
    c. cyclic
    d. directional control system
    e. None of the above

4. What is the flight attitude?
    a. the environment immediately around the plane
    b. the morale of the flight crew
    c. the position of the plane in motion
    d. the inclination of the elevators
    e. the positions of the ailerons

5. The curvature of an airfoil is known as the
    a. bank.
    b. position.
    c. camber.
    d. angle.
    e. attitude.

6. Which of the following is considered one of the primary flight controls?
    a. elevators
    b. leading edge devices
    c. flaps
    d. spoilers
    e. trim tabs

7. What is the term in aviation for movement around the plane's longitudinal axis?
   a. stalling
   b. rolling
   c. pitching
   d. leaning
   e. yawing

8. What is the primary determinant of air pressure in the flight envelope?
   a. the altitude at which the plane is flying
   b. the camber of the wings
   c. the amount of lift that the airfoils can create
   d. the pitch of the elevators
   e. the humidity of the air

9. A plane is said to have conventional landing gear when the third wheel is
   a. aligned with the second wheel.
   b. under the tail.
   c. directly behind the first wheel.
   d. under the nose.
   e. underneath the cockpit.

10. Which of the following is NOT part of the empennage?
   a. rudder
   b. trim tab
   c. elevator
   d. aileron
   e. horizontal stabilizer

11. A glider would be most likely to have a
   a. delta wing.
   b. triangular wing.
   c. forward swept wing.
   d. backward swept wing.
   e. straight wing.

12. Which of the following is NOT considered part of a plane's basic weight?
   a. crew
   b. fuel
   c. external equipment
   d. internal equipment
   e. fuselage

13. The support structure that runs the length of the fuselage in a monocoque plane is called a
   a. counter.
   b. former.
   c. truss.
   d. stringer.
   e. bulkhead.

14. Which control is used to manipulate the elevators on an airplane?
   a. throttle
   b. rudder
   c. joystick
   d. pedals
   e. collective

15. Which maneuver is appropriate when a pilot needs to descend quickly onto a shorter-than-normal runway?
   a. descent at minimum safe airspeed
   b. idle
   c. partial power descent
   d. glide
   e. stall

16. In aviation, which of the following is NOT one of the possible functions of a spoiler?
   a. diminishing lift
   b. raising the nose
   c. increasing drag
   d. reducing adverse yaw
   e. enabling descent without speed reduction

17. The vertical axis of a plane extends upward through the plane's
   a. cockpit.
   b. tail.
   c. landing gear.
   d. center of mass.
   e. geometric center.

18. When is trimming necessary?
   a. when the plane is ascending or descending
   b. after the elevators have been deflected upwards
   c. after the ailerons have been adjusted
   d. after the elevators have been deflected downwards
   e. after any change in the flight condition

19. How far away is an approaching plane when the Runway Centerline Lighting System lights become solid red?
   a. Five hundred feet
   b. One thousand feet
   c. Three thousand feet
   d. Five thousand feet
   e. One mile

20. What is the Coriolis force?
   a. the extra lift generated by a helicopter once it has exited its own downwash
   b. the force that spins the rotors of a helicopter even when there is no power from the engine
   c. the greater downwash at the rear half of the rotor disc, as compared with the front
   d. the phenomenon in which the effects of a force applied to a spinning disc occur ninety degrees later
   e. the change in rotational speed caused by the shift of the weight towards or away from the center of the spinning object

Answer Key

Verbal Analogies

1. B: Rash. Chastise and reprimand are synonyms. The answer choice synonym for impetuous is rash.

2. C: Femur. The humerus is a bone in the arm; the femur is a bone in the leg.

3. A: Quotient. A quotient is the result of division as a product is the result of multiplication.

4. C: Asparagus. As a sweater is something one wears, asparagus is something one eats.

5. A: Famished. As a person who is impecunious needs money, so a person who is famished needs food.

6. A: Protest. Denigrate and malign are synonyms. The answer choice synonym for demur is protest.

7. D: Generous. Obeisance and deference are synonyms. The answer choice synonym for munificent is generous.

8. D: Sow. A female goat is a nanny and a female pig is a sow.

9. B: Paucity. Cache and reserve are synonyms. The answer choice synonym for dearth is paucity.

10. E: Refuge. Arable and farmable are synonyms. The answer choice synonym for asylum is refuge.

11. A: Peripatetic. Myriad and few are antonyms. The answer choice antonym for stationary is peripatetic.

12. A: Flagon. As a mansion is a large house, so a flagon is a large bottle.

13. B: Synonyms. As a dictionary is a collection of definitions, so a thesaurus is a collection of synonyms.

14. B: Stubborn. Abstruse and esoteric are synonyms. The answer choice synonym for adamant is stubborn.

15. C: Herd. A group of bees is called a hive and a group of cattle is called a herd.

16. E: Meridian. A parallel is a line of latitude, while a meridian is a line of longitude.

17. B: Provocation. Prevention and deterrence are synonyms. The answer choice synonym for incitement is provocation.

18. A: Gauge. Value and worth are synonyms. The answer choice synonym for measure is gauge.

19. A: Oppose. Enervate and energize are antonyms. The answer choice antonym for espouse is oppose.

20. D: The analogy is tool to worker. A carpenter uses a hammer just as a doctor uses a stethoscope.

21. B: The analogy describes characteristic location. An armoire is usually kept in a bedroom, just as a desk is usually kept in an office.

22. C: This analogy describes a food source to animal relationship. Just as plankton is a food source for whales, so is bamboo a food source for pandas.

23. C: The analogy is geographic location. Just as the tundra is located in the Arctic regions, so are savannas located in tropic regions.

24. A: This is an analogy of relative degree. *Parched* is a more intense degree of *thirsty,* just as *famished* is a more intense degree of *hungry.*

25. E: This is an analogy based on antonyms. *Felicity,* or happiness, is the opposite of *sadness,* just as *ignominy,* or disgrace, is the opposite of *honor.*

Arithmetic Reasoning

1. B: First, find the total before taxes: $7.50 + $3.00 = $10.50
Then, calculate 6% of the total: $10.50 * .06 = $0.63
Finally, add the tax to find the total cost: $10.50 + $0.63 = $11.13

2. D: There are 7 days in a week. Knowing that the chef can make 25 pastries in a day, the weekly number can be calculated:
25 * 7 = 175

3. B: The woman has four days to earn $250. To find the amount she must earn each day, divide the amount she must earn ($250) by 4:
$250 / 4 = $62.50

4. A: To find the number of cars remaining, subtract the number of cars that were sold from the original number: 476 – 36 = 440

5. B: Calculate 0.5% of $450: $450 * 0.005 = $2.25
This is the amount of interest she will earn.

6. C: First, figure out how much the second child contributed: $24.00 - $15.00 = $9.00
Then, calculate how much the first two children contributed in total: 24 + 9 = $33.00
Finally, figure out how much the third child will have to contribute:
$78.00 - $33.00 = $45.00

7. C: First, figure out how many points the first woman will earn: 3 * 5 = 15
Then, figure out how many points the second woman will earn: 6 * 5 = 30
Then, add these two values together: 30 + 15 = 45 points total.

8. E: First, calculate 13% of 540 = 70
Then, add this value onto the original number of workers: 540 + 70 = 610
610 is the number of people that the company will employ after the expansion.

9. D: To find the number of apartments on each floor, divide the total number of apartments by the number of floors:
65 / 13 = 5

10. A: First, find the total number of pens: 5 * 3 = 15
Then, find the total number of pencils: 3 * 7 = 21
Finally, express it as a ratio 15 : 21

11. C: To calculate his new salary, add his raise to his original salary:
$15.23 + $2.34 = $17.57

12. A: To find the total number of passengers, multiply the number of planes by the number of passengers each can hold: 6 * 300 = 1800

13. B: Currently, there are two men for every woman. If the number of women is doubled (1 * 2 = 2), then the new ratio is 2:2. This is equivalent to 1:1.

14. C: First, calculate 3% of 250 pounds: 250 * 0.03 = 7.5 pounds
Calculate how much she weighs at the end of the first week: 250 – 7.5 = 242.5 pounds
Calculate 2% of 242.5: 242.5 * 0.02 = 4.85 pounds
Add the two values together to get the total: 7.5 + 4.85 = 12.35

15. D: Divide the total distance she must travel (583km) by the number of kilometers she drives each hour (78km) to figure out how many hours it will take to reach her destination:
583 km / 78 km = 7.47 hours

16. D: One gallon of paint can paint three rooms, so to find out how many 28 gallons can do, that number must be multiplied by 3: 28 * 3 = 84 rooms

17. A: Each earns $135, so to find the total earned, that amount must be multiplied by the number of workers: $135 * 5 = $675

18. C: First, calculate her score on the second test: 99 – 15 = 84
Then, calculate her score on the third test: 84 + 5 = 89

19. A: To find out how much he has remaining, both numbers must be subtracted from the original amount ($50.00): $50.00 - $15.64 - $7.12 = $27.24

20. E: Divide the number of students (600) by the number of classrooms they will share (20):
600 / 20 = 30

21. C: To calculate this value, divide the number of dogs (48) by the number of workers that are available to care for them (4):
48 / 4 = 12

22. C: First, calculate the length of the second office: 20 + 6 = 26 feet
Then, add both values together to get a combined length: 26 + 20 = 46 feet

23. C: Find the total cost of the items: $6.66 + $159.23 = $165.89
Then, calculate how much each individual will owe: $165.89 / 4 = $41.47

24. A: To answer this question, simply calculate half of 140 acres: 140 / 2 = 70 acres

25. C: First, calculate how many he has after selling 45: 360 – 45 = 315
Then, calculate how many he has after buying 85: 315 + 85 = 400

Word Knowledge

1. A: Spoiled has a number of meanings, and one of them is ruined. If you said somebody spoiled your fun, it would convey the same meaning as saying somebody ruined your fun.

2. B: An oath is a promise. For example, if you make an oath to keep a secret, you are promising to keep that secret.

3. B: When you inquire about something, you are asking about it or requesting more information. For example, if you told somebody you inquired about a job, it would mean you asked about it.

4. C: If you say that you comprehend something, it is the same as saying you understand it. For example, saying you comprehend what another person is saying is the same as saying you understand them.

5. A: To say that something is apparent implies that it is clear or obvious. For example, saying that it is apparent that somebody wants a job is the same as saying it is clear they want the job.

6. C: Silent or silence indicates quiet and calm. To enjoy the silence of the night is to enjoy the complete quiet of the night.

7. E: Absolutely, when used to describe a feeling or state of mind, means completely or totally. For example, saying you are absolutely certain that you made the right decision or saying you are completely certain you made the right decision conveys the same meaning.

8. D: Something that has been modified has been changed. Saying you modified your plans or saying that you changed them conveys the same meaning.

9. A: Something that is delicate can also be described as fragile. Saying that a crystal figurine is delicate or saying it is fragile conveys the same meaning.

10. B: Festivities are often commonly known as celebrations. Attending festivities implies that you are attending a celebration or party.

11. B: To say that someone is exhausted or to say that they are tired conveys a similar meaning. Usually, exhausted is a word used to describe extreme tiredness.

12. B: To cleanse something is to clean or wash it. Saying you cleansed your face or clothes is the same as saying you washed them.

13. A: To battle something is to fight it. To say that two armies battled each other and to say they fought each other conveys the same meaning.

14. C: To wander is to roam. To say someone wandered around a mall is to say they roamed or walked around aimlessly, without a specific goal or destination in mind.

15. E: Something that is done abruptly is done suddenly and without warning. For example, saying the car stopped abruptly and saying it stopped suddenly conveys the same meaning.

16. A: Somebody who has been tricked has been conned. To trick somebody is to con them, which implies that dishonest methods are used to convince another to do something they wouldn't normally do.

17. C: When used as an adjective extremely has the same meaning as very. Saying somebody is extremely happy and saying they are very happy conveys the same meaning.

18. A: To have doubts is to have uncertainties or hesitations. To say that someone is doubtful about something means that they are uncertain.

19. D: To describe something as peculiar is to say it is strange or out of the ordinary. For example, saying you are in a strange situation or saying you are in a peculiar situation conveys the same meaning.

20. B: Describing somebody as courteous implies that they are polite and well-mannered. Polite and courteous both convey the same meaning.

21. C: When somebody says they are troubled by something, it means that they are bothered by it.

22. A: Perspiration is another word for sweat. Saying somebody is perspiring is the same as saying they are sweating.

23. B: Tremble is another word for shake. To say somebody or something trembled means that it shook or shuddered.

24. A: Adhered is often used as another word for stuck. For example, to say a piece of tape adhered to the wall conveys the same meaning as saying the piece of tape stuck to the wall.

25. D: When something is described as tidy, it usually means that it is neat and that things are in their proper place. Saying a house is tidy and saying it is neat conveys the same meaning.

Math Knowledge

1. B: The perimeter of a figure is the sum of all of its sides. Since a rectangle's width and length will be the same on opposite sides, the perimeter of a rectangle can be calculated by using the following formula: perimeter = 2(width) + 2(length)
Using the numbers given in the question:
perimeter = 2(7cm) + 2(9cm)
perimeter = 14cm + 18cm
perimeter = 32cm

2. D: First, gather the like terms on opposite sides of the equation to make it easier to solve:
$-3q - 4q \geq -30 - 12$
$-7q \geq -42$
Then, divide both sides by -7 to solve for q:
$-7q/-7 \geq -42/-7$
$q \geq 6$
Finally, when both sides are divided by a negative number, the direction of the sign must be reversed:
$q \leq 6$

3. C: To solve for x, it is necessary to add 6 to both sides to isolate the variable:
$x - 6 + 6 = 0 + 6$
$x = 6$

4. A: To calculate the value of this expression, substitute -3 for x each time it appears in the expression: $3(-3)^3 + (3(-3) + 4) - 2(-3)^2$
According to the order of operations, any operations inside of brackets must be done first:
$3(-3)^3 + (-9 + 4) - 2(-3)^2$
$3(-3)^3 + -5 - 2(-3)^2$
Then, the value of the expression can be calculated:
$3(-27) + -5 - 2(9)$
$-81 + -5 - 18$
-104

5. C: First, combine like terms to make the equation easier to solve:
$3x + 2x = 45 + 30$
$5x = 75$
Then, divide both sides by 5 to solve for x:
$5x/5 = 75/5$
$x = 15$

6. A: First, add 25 to both sides to isolate x:
$1/4x - 25 + 25 \geq 75 + 25$
$1/4x \geq 100$
Then, multiply both sides by 4 to solve for x:
$1/4x * 4 \geq 100 * 4$
$x \geq 400$

7. A: First, add 5 to both sides to isolate x:
$x^2 - 5 + 5 = 20 + 5$
$x^2 = 25$
Then, take the square root of both sides to solve for x
$\sqrt{x^2} = \sqrt{25}$
$x = 5$

8. B: First, we must calculate the length of one side of the square. Since we know the perimeter is 8cm, and that a square has 4 equal sides, the length of each side can be calculated by dividing the perimeter (8cm) by 4: 8cm / 4 = 2cm
The formula for the area of a square is length^2
Therefore, to calculate the area of this square: $2cm^2$ or 2cm * 2cm
Area = $4cm^2$

9. D: To find the value of this expression, substitute the given values for x and y into the expression:
3(4)(2) – 12(2) + 5(4)
Then, calculate the value of the expression:
3*8 – 12*2 + 5*4
24 – 24 + 20
20

10. D: First, subtract 10 from both sides to isolate x:
0.65x + 10 – 10 = 15 – 10
0.65x = 5
Then, divide both sides by 0.65 to solve for x:
0.65x/0.65 = 5/0.65
x = 7.69

11. B: Use the FOIL method (first, outside, inside, and last) to get rid of the brackets:
$12x^2 -18x + 20x -30$
Then, combine like terms to simplify the expression:
$12x^2 -18x + 20x -30$
$12x^2 + 2x -30$

12. B: To simplify this expression, it is necessary to follow the law of exponents that states: $x^n/x^m = x^{n-m}$
First, the 50 can be divided by 5: 50/5 = 10
Then, it is simply a matter of using the law of exponents described above to simplify the expression:
$10x^{18-5}t^{6-2}w^{3-2}z^{20-19}$
$10x^{13}t^4wz$

13. E: To calculate the value of this permutation, it is necessary to multiply each number between one and 4: 1 * 2 * 3 * 4 = 24

14. D: Because it is a cube, it is known that the width and the height of the cube is also 5cm. Therefore, to find the volume of the cube, we must cube 5cm: $5cm^3$
This is the same as: 5 * 5 * 5 = 125
The volume of the cube is $125cm^3$.

15. A: First, factor this equation to make solving for x easier:
$(x - 6)(x - 7) = 0$
Then, solve for both values of x:
1) $x - 6 = 0$
$x = 6$
2) $x - 7 = 0$
$x = 7$

16. C: The area of a triangle can be calculated by using the following formula: A = 1/2b*h
Therefore, by using the values given in the question:
A = 1/2(12cm) * 12cm
A = 6cm * 12cm
A = $72cm^2$

17. D: To simplify this expression, it is necessary to observe the law of exponents that states:
$x^n * x^m = x^{n+m}$
Therefore: $3*7x^{7+2} + 2*9y^{12+3}$
$21x^9 + 18y^{15}$

18. C: First, subtract 27 from both sides to isolate x:
x/3 + 27 – 27 = 30 – 27
x/3 = 3
Then, both sides must be multiplied by 3 to solve for x:
$\not{3}(x/\not{3}) = 3 * 3$
x = 9

19. B: To find the slope of a line, it is necessary to calculate the change in y and the change in x:
Change in y: 1 – 8 = -7
Change in x: 4 – (-13) = 17
The slope of a line is expressed as change in y over change in x: -7/17

20. A: To solve for x, it is necessary to calculate the value of 20% of 200:
200 * 0.20 = 40
Therefore, x = 40

21. B: First, calculate the total number of balloons in the bag: 47 + 5 + 10 = 62
Ten of these are black, so divide this number by 62, then multiply by 100 to express the probability as a percentage:
10 / 62 = 0.16
0.16 * 100 = 16%

22. B: First, it is easier to find out how many tickets are sold for one winner.
If there are 2 winners for every 100 tickets, there is 1 winner for every 50 tickets.
If ten tickets are bought, the chances of winning are 10 in 50.
This can also be expressed as 1 in 5.

23. E: To find the volume of a rectangular prism, the formula is length * width * height.
Therefore, for this rectangular prism, volume = 10cm * 5cm * 6cm
The volume of this rectangular prism is $300cm^3$

24. C: To calculate the midpoint of a line, find the sum of the points and divide by two.
For x, the midpoint can be calculated as follows: $6 + 10 = 16$; $16/2 = 8$
For y, the midpoint can be calculated as follows: $40 + 20 = 60$; $60/2 = 30$
Therefore, the midpoint is (8, 30)

25. A: First, subtract 60 from both sides to isolate x:
$5x + 60 - 60 = 75 - 60$
$5x = 15$
Then, divide both sides by 5 to solve for x:
$5x/5 = 15/5$
$x = 3$

Reading Comprehension

1. A. Explaining the qualities of air that may affect flight is the primary purpose of the passage.

2. C. The best definition for *inversely* as it is used in the second paragraph is *in the opposite direction*. The author is indicating that the density of air decreases as the temperature rises, and increases as the temperature falls.

3. D. A pilot can expect the air density to decrease as the plane gains altitude. At the end of the second paragraph, the author states that a gain in altitude will usually lead to a decrease in air density, no matter what changes there may be in the temperature.

4. B. The author would most likely agree that air density is more important than relative humidity. Though the passage states that humidity can have an effect on aircraft performance, the author admits that it is not considered an essential factor. Indeed, humidity is just one of three factors (along with pressure and temperature) that affect air density.

5. C. The most likely reason why there is no chart for assessing the effects of humidity on density altitude is that humidity does not affect flight performance very much. The author mentions several times that flight performance is not significantly affected by humidity, and so it is seems that a special chart for this purpose would be unnecessary.

6. D. *Influences on Climb Performance* would be the best title for this passage. The passage surveys the various factors that affect climb performance and the choices made by pilots as they gain altitude, both during take-off and while the aircraft is already in flight.

7. E. The best definition for *pronounced* as it is used in the second paragraph is *noticeable*. The author is stating that the weight of an aircraft has a significant impact on aircraft performance. In this sentence *pronounced* is being used as an adjective, but it can also be used as a verb, meaning *spoke* or *said*.

8. C. An increase in weight means that the angle of attack must be higher in order to maintain altitude. Greater weight diminishes reserve power, lowers the climb rate, diminishes the maximum rate of climb, and increases drag.

9. B. The author would most likely agree that at the end of a long journey a plane will have a higher maximum rate of climb. The maximum rate of climb of a plane increases as the weight decreases, and the weight of a plane will decrease as it burns off fuel over the course of a long journey.

10. A. A helicopter that weighs two tons and has rotor blades that cover five hundred square feet would have a disc loading measure of four pounds per square foot. Pounds per square foot is the standard measure for disc loading. There are two thousand pounds in a ton. Disc loading is calculated by dividing the weight of the helicopter by the area covered by the rotor blades.

11. B. Discussing the interrelationships of airspeed, power, and pitch attitude is the primary purpose of the passage. Answer choices *C* and *D* are partially correct, but they leave out large sections of the passage, and therefore do not comprehensively describe the purpose or the content of the passage.

12. A. *Inadvertently*, as it is used in the fifth paragraph, most nearly means *unintentionally*. The author is indicating that a pilot can enter the region of reversed command without meaning to, if he or she attempts to climb out of ground effect without first attaining normal climb pitch attitude and airspeed.

13. B. Most flight occurs in the region of normal command. In the region of normal command, increasing power increases the airspeed, and decreasing power decreases the airspeed. This information is given in the last sentence of the second paragraph.

14. E. The author would most likely agree that as the speed of flight decreases, the power required to maintain altitude increases. This inverse relationship between required power and airspeed is expressed in the last sentence of the first paragraph.

15. D. The best title for this passage would be *Power Requirements During Flight*. The passage discusses how the need for and effects of changes in power are influenced by factors such as airspeed, pitch attitude, and altitude.

16. E. The primary purpose of the passage is to describe the factors that influence landing distance. The passage begins by discussing the minimum and normal landing distances, and goes on to cover influences on landing distance, as for instance gross weight, wind, and density altitude.

17. B. When making a normal landing, a pilot will rely on aerodynamic drag in order to avoid wearing down the brakes and tires. This point is made several times in the third paragraph of the passage.

18. A. The best definition for *principal* as it is used in the fourth paragraph is *most important*. The author is trying to make the point that gross weight has an enormous effect on landing distance.

19. C. A heavier plane must be landed at a higher airspeed to avoid hitting the runway with too much force. This idea is explored in the fourth paragraph. In order to generate an amount of lift sufficient for a smooth landing, a higher airspeed must be maintained.

20. C. The author would most likely agree that gross weight and minimum landing distance are positively correlated. The information in the passage makes it clear that as gross weight rises, minimum landing distance increases.

21. D. The primary purpose of the article is to discuss aeronautical decision making, or ADM. The article does give some examples of decision-making strategies, but these are given in the context of a description of ADM, not as the main body of the article.

22. D. The passage explains that aviation safety is distinct from other areas of safety because there is a much smaller margin for error. In other words, even small accidents in aviation can be catastrophic, because of the risks inherent in flight.

23. A. The closest definition for *conjunction* as it is used in the second paragraph is *combination*. The author is stating that the FAA manuals worked well when combined with the usual flight training.

24. A. The author would most likely agree that the body of knowledge about ADM is increasing, and this will have a positive effect on flight safety. The article details the efforts to improve ADM, and suggests that these have already improved flight safety a great deal.

25. E. For a pilot, reading the account of a recent aviation accident is an opportunity for an indirect learning experience. The passage distinguishes between direct learning experiences, which are events in one's own life, and indirect learning experiences, which are things that happen to others.

Situational Judgment

1. C (Most effective)
2. A (Least effective)
3. A (Most effective)
4. C (Least effective)
5. E (Most effective)
6. B (Least effective)
7. D (Most effective)
8. A (Least effective)
9. E (Most effective)
10. C (Least effective)
11. B (Most effective)
12. C (Least effective)
13. A (Most effective)
14. B (Least effective)
15. E (Most effective)
16. B (Least effective)
17. C (Most effective)
18. A (Least effective)
19. A (Most effective)
20. B (Least effective)
21. B (Most effective)
22. D (Least effective)
23. A (Most effective)
24. C (Least effective)
25. B (Most effective)
26. E (Least effective)
27. D (Most effective)
28. A (Least effective)
29. B (Most effective)
30. C (Least effective)
31. E (Most effective)
32. B (Least effective)
33. A (Most effective)
34. D (Least effective)
35. D (Most effective)
36. C (Least effective)
37. A (Most effective)
38. D (Least effective)
39. E (Most effective)
40. A (Least effective)
41. B (Most effective)
42. C (Least effective)
43. E (Most effective)
44. A or C (Least effective)
45. A (Most effective)
46. D (Least effective)
47. A (Most effective)
48. B (Least effective)
49. B (Most effective)
50. D (Least effective)

Physical Science

1. B: A long nail or other type of metal, substance or matter that is heated at one end and then the other end becomes equally hot is an example of conduction. Conduction is energy transfer by neighboring molecules from an area of hotter temperature to cooler temperature.

2. E: The measure of energy within a system is called heat.

3. B: They have a different number of neutrons. The distinguishing feature of an isotope is its number of neutrons. Two different isotopes of the same element will have the same number of protons but different numbers of neutrons.

4. A: Fission is a nuclear process where atomic nuclei split apart to form smaller nuclei. Nuclear fission can release large amounts of energy, emit gamma rays and form daughter products. It is used in nuclear power plants and bombs.

5. D: The process whereby a radioactive element releases energy slowly over a long period of time to lower its energy and become more stable is best described as decay. The nucleus undergoing decay spontaneously releases energy, most commonly through the emission of an alpha particle, a beta particle or a gamma ray.

6. B: Light within a single medium travels in a straight line. When it changes to a different medium, however, the light rays bend according to the refractive index of each substance. Light coming from the submerged portion of the pencil is refracted as it passes through the air-water barrier, giving the perception of a bent pencil.

7. A: Hertz (Hz) is a unit of measure used for frequency, often described as 1 cycle/second. In the context of wave motion, it is the number of complete waves that pass a given point in one second.

8. B: A cyclist coasting up a hill is trading his speed for increased altitude. This is an example of kinetic energy being converted to potential energy. The other options are examples of potential energy being converted to kinetic, kinetic energy being dissipated, and conservation of kinetic energy.

9. E: Phase changes such as boiling, melting, and freezing are physical changes. No chemical reaction takes place when water is boiled.

10. A: The center of an atom is known as the nucleus. It is composed of protons and neutrons.

11. A: Sublimation is the process of a solid changing directly into a gas without entering the liquid phase.

12. D: Mendeleev was able to connect the trends of the different elements behaviors and develop a table that showed the periodicity of the elements and their relationship to each other.

13. A: Density is mass per unit volume, typically expressed in units such as g/cm^3, or kg/m^3.

14. C: The closer the data points are to each other, the more precise the data. This does not mean the data is accurate, but rather that the results are reproducible.

15. E: Current is measured in units of amperes or amps.

16. C: In order for a solar eclipse to occur, the moon must come directly between the earth and the sun, blocking the sun's light from the earth.

17. B: Inertia is the tendency of objects that are in motion to continue moving in the same direction. The turning car initiates a change in direction, but the passengers' mass wants to continue going straight, causing them to feel a pull in that direction. This phenomenon is sometimes referred to as centrifugal force.

18. B: According to the ideal gas law, when volume is held constant, the temperature of a gas is directly proportional to the pressure of the gas. Thus, when the temperature is increased, the pressure will also increase.

19. B: The amplitude of a sound wave is what determines how loud the sound is perceived by the ear.

20. D: In all of the other examples, there is a person applying work to the book, either directly, by being picked up, pushed, or thrown, or indirectly, by being carried in a backpack. In the example of a book being released so that it falls, the only work being applied to the book is being done by gravity.

Table Reading

1. E
2. D
3. A
4. A
5. E
6. D
7. A
8. C
9. A
10. C
11. E
12. E
13. A
14. D
15. A
16. B
17. C
18. A
19. C
20. B
21. C
22. B
23. E
24. A
25. C
26. C
27. E
28. B
29. C
30. D
31. A
32. C
33. C
34. E
35. A
36. C
37. A
38. A
39. C
40. D

Instrument Comprehension

1. D

2. B

3. B

4. C

5. D

6. A

7. C

8. B

9. C

10. B

11. C

12. C

Block Counting

1. 6: 1 on the front, 2 on the back, 2 on the top, 1 on the bottom.

2. 3: 2 on the right, 1 on the top.

3. 6: 1 on the front, 3 on the left, 2 on the top.

4. 5: 1 on the back, 1 on the right, 3 on the top.

5. 5: 3 on the front, 2 on the top.

6. 9: 1 on the front, 3 on the back, 3 on the top, 2 on the bottom.

7. 4: 1 on the front, 1 on the back, 1 on the right, 1 on the bottom.

8. 3: 1 on the right, 2 on the top.

9. 5: 2 on the front, 1 on the back, 1 on the left, 1 on the bottom.

10. 9: 4 on the front, 4 on the back, 1 on the top.

11. 6: 2 on the back, 3 on the left, 1 on the bottom.

12. 4: 1 on the right, 3 on the top.

Aviation Information

1. C: A Visual Approach Slope Indicator should be visible from approximately twenty miles away at night. A Visual Approach Slope Indicator (VASI) is a common feature at large airports. This system helps guide the approaching pilot to the runway. The pilot will see white lights at the lower border of the glide path, and red lights at the upper border. In normal conditions, the lights of the VASI system should be visible for three to five miles during the day, and for twenty miles at night. If the VASI system is working properly, the plane will be safe so long as it stays within ten degrees of the extended runway centerline and four nautical miles of the runway threshold.

2. E: Torque is not one of the forces a pilot must manage during flight. The four forces a pilot must manage are lift, gravity, thrust, and drag. Lift pushes the plane up, gravity pulls the plane down, thrust propels the plane forward, and drag holds the plane back. The overall admixture of these forces as they operate on the plane is called the flight envelope.

3. A: The collective affects the angle of the main rotor blades of a helicopter. It is a long tube that extends from the floor of the cockpit. In most helicopters, it is situated on the pilot's left. The collective has two parts: a handle that can be raised or lowered, to control the pitch of the blades; and a throttle, to control the torque of the engine. The handle is the part that affects the angle of the main rotor blades. When it is raised, the leading edge of the blade is raised higher than the trailing edge.

4. C: The flight attitude is the position of a plane in motion. The flight attitude is described in terms of its position with respect to three axes: vertical, lateral, and longitudinal. The vertical axis runs up through the plane's center of gravity. A plane's position with respect to this axis is known as its yaw. The lateral axis of a plane runs from wingtip to wingtip, and the motion of the plane around this axis is known as pitch. The longitudinal axis, finally, is an imaginary line extending from the nose of the plane to its tail. The position of the plane in relation to the longitudinal axis is called roll. The attitude of the plane is controlled with the joystick, rudder pedals, and throttle.

5. C: The curvature of an airfoil is known as the camber. An airfoil (wing) is considered to have a high camber if it is very curved. A related piece of wing terminology is the mean camber line, which runs along the inside of the wing, such that the upper and lower wings are equal in thickness.

6. A: The elevators are considered one of the primary flight controls. The elevators are responsible for the plane's pitch, or movement around the lateral axis. The other primary flight controls are the ailerons and the rudder. The ailerons control the roll, or movement around the longitudinal axis, while the rudder controls the yaw, or movement around the vertical axis. The secondary flight controls are the flaps, spoilers, leading edge devices, and trim systems.

7. B: In aviation, the term for movement around the plane's longitudinal axis is rolling. The longitudinal axis runs from the nose of the plane to its tail. For the most part, the plane's roll is controlled by the joystick. By moving the stick to the right or left, the pilot dips the wings.

8. A: The altitude at which the plane is flying is the primary determinant of air pressure in the flight envelope. The most important elements of the flight envelope are the temperature, air pressure, and humidity. The conditions in the atmosphere have a great deal of influence over the amount of lift created by the airfoils. Greater air pressure is the same as greater air density. A plane will generate greater lift when it is in cool air, because cool air is less dense than warm air.

9. B: A plane is said to have conventional landing gear when the third wheel is under the tail. Typically, a plane's landing gear will consist of three wheels or sets of wheels. Two of these are under either wing or on opposing sides of the fuselage. In the conventional arrangement, the third wheel or wheel set is under the tail, while in the tricycle arrangement it is under the nose. This third wheel can rotate, which will make it possible for the plane to turn while moving on the ground.

10. D: An aileron is not part of the empennage. The empennage, otherwise known as the tail assembly, includes the elevators, vertical and horizontal stabilizers, rudders, and trim tabs. A fixed wing aircraft will typically have both vertical and horizontal stabilizers, which are immobile surfaces that extend from the back of the fuselage. The horizontal stabilizers have mobile surfaces along their trailing edges; these surfaces are called the elevators. The elevators deflect up and down to raise or lower the nose of the plane. The rudder is a single, large flap connected by a hinge to the vertical stabilizer. The rudders back-and-forth motion controls the motion of the plane with respect to its vertical axis. The trim tabs, finally, are connected to the trailing edges of one or more of the primary flight controls (i.e., ailerons, elevators, rudder).

11. E: A glider would be most likely to have a straight wing. A straight wing may be tapered, elliptical, or rectangular. This planform (wing shape) is common in aircraft that move at extremely low speed. A straight wing is often found on sailplanes and gliders. The swept wing, on the other hand, is appropriate for high-speed aircraft. The wing may be swept forward or back. This will make the plane unstable at low speeds, but will produce much less drag. A swept wing requires high-speed takeoff and landing. The delta wing, which is also known as the triangular wing, has a straight trailing edge and a high angle of sweep. This allows the plane to take off and land at high speeds.

12. A: The crew is not considered part of a plane's basic weight. The basic weight is the aircraft plus whatever internal or external equipment will remain on the plane during its journey. The crew is not included in the basic weight, though it is a part of the operating weight (basic weight plus crew), gross weight (total weight of the aircraft at any particular time), landing gross weight (weight of the plane and its contents upon touchdown), and zero fuel weight (weight of the airplane when it has no usable fuel).

13. D: The support structure that runs the length of the fuselage in a monocoque plane is called a stringer. In a monocoque plane, the fuselage is supported by stringers, formers, and bulkheads. Stringers and formers are generally made out of the same material, though they run perpendicular to one another (that is, formers run in circles around the width of the fuselage). Bulkheads are the walls that divide the sections of the fuselage. The other style of fuselage, known as a truss, is composed of triangular groupings of aluminum or steel tubing.

14. C: The joystick is used to manipulate the elevators. When the stick is pulled back, the elevators deflect upwards. This decreases the camber of the horizontal tail surface, which moves the nose up and pushes the tail down. Pushing the joystick forward, on the other hand, pushes the elevators down, which creates an upward force on the tail by increasing the camber of the horizontal tail surface.

15. A: When a pilot needs to descend quickly onto a shorter-than-normal runway, he or she will descend at the minimum safe airspeed. A descent at the minimum safe airspeed is achieved by slightly lifting the nose and moving the plane into the landing configuration. During such a descent, the plane should not exceed 1.3 times the stall speed. This technique is appropriate for landing

quickly on a short runway because the rate of descent is much faster. However, if the rate of descent should become too great, the pilot should be ready to increase power.

16. B: Raising the nose is not one of the possible functions of a spoiler. Spoilers can diminish lift and reduce drag, which enables the plane to descend without reducing its speed. However, spoilers also control the plane's roll, partly by reducing any adverse yaw. A pilot uses the spoiler in this way by raising the spoiler on the side of the turn. That side will thereby have less lift and more drag, making it drop. The plane will then bank and yaw in the intended direction. When the pilot raises both of the spoilers at the same time, the plane will descend without losing any speed. Another incidental benefit of spoilers is improved brake performance, which occurs because the plane has lift and is pushed down towards the ground.

17. D: The vertical axis of a plane extends up through the plane's center of gravity. Movement around this axis is called yawing. The position of a plane is also described with respect to the lateral and longitudinal axes. The lateral axis extends from wingtip to wingtip. The longitudinal axis runs from the nose of the plane to its tail.

18. E: Trimming is necessary after any change in the flight condition. Trimming is the adjustment of the trim tabs, which are small flaps that extend from the trailing edges of the elevators, rudder, and ailerons. Trimming generally occurs after the pilot has achieved the desired pitch, power, attitude, and configuration. The trim tabs are then used to resolve the remaining control pressures. A small plane may only have a single tab, which is controlled with a small wheel or crank.

19. D: When the Runway Centerline Lighting System lights become solid red, an approaching plane is one thousand feet away. A Runway Centerline Lighting System is a line of white lights every fifty feet or so along the centerline. The lights change their color and pattern as the plane nears the runway. When the plane gets within 3000 feet of the runway, the lights blink red and white. Within a thousand feet, the lights will turn solid red.

20. E: The Coriolis force is the change in rotational speed caused by the shift of the weight towards or away from the center of the spinning object. This phenomenon has an important application for helicopters, in which the rotor will move faster or will require less power to maintain its speed when the weight is closer to the base of the blade. The other answer choices are similarly related to helicopters. The extra lift generated by a helicopter once it has exited its own downwash is known as translational lift. The force that spins the rotors of a helicopter even when there is no power from the engine is autorotation. Greater downwash at the rear half of the rotor disc, as compared to the front half, is the result of applying the lateral cyclic. The phenomenon in which the effects of a force applied to a spinning disc occur ninety degrees later is called gyroscopic precession.

Secret Key #1 - Time is Your Greatest Enemy

Pace Yourself

Wear a watch. At the beginning of the test, check the time (or start a chronometer on your watch to count the minutes), and check the time after every few questions to make sure you are "on schedule."

If you are forced to speed up, do it efficiently. Usually one or more answer choices can be eliminated without too much difficulty. Above all, don't panic. Don't speed up and just begin guessing at random choices. By pacing yourself, and continually monitoring your progress against your watch, you will always know exactly how far ahead or behind you are with your available time. If you find that you are one minute behind on the test, don't skip one question without spending any time on it, just to catch back up. Take 15 fewer seconds on the next four questions, and after four questions you'll have caught back up. Once you catch back up, you can continue working each problem at your normal pace.

Furthermore, don't dwell on the problems that you were rushed on. If a problem was taking up too much time and you made a hurried guess, it must be difficult. The difficult questions are the ones you are most likely to miss anyway, so it isn't a big loss. It is better to end with more time than you need than to run out of time.

Lastly, sometimes it is beneficial to slow down if you are constantly getting ahead of time. You are always more likely to catch a careless mistake by working more slowly than quickly, and among very high-scoring test takers (those who are likely to have lots of time left over), careless errors affect the score more than mastery of material.

Secret Key #2 - Guessing is not Guesswork

You probably know that guessing is a good idea. Unlike other standardized tests, there is no penalty for getting a wrong answer. Even if you have no idea about a question, you still have a 20-25% chance of getting it right.

Most test takers do not understand the impact that proper guessing can have on their score. Unless you score extremely high, guessing will significantly contribute to your final score.

Monkeys Take the Test

What most test takers don't realize is that to insure that 20-25% chance, you have to guess randomly. If you put 20 monkeys in a room to take this test, assuming they answered once per question and behaved themselves, on average they would get 20-25% of the questions correct. Put 20 test takers in the room, and the average will be much lower among guessed questions. Why?

1. The test writers intentionally write deceptive answer choices that "look" right. A test taker has no idea about a question, so he picks the "best looking" answer, which is often wrong. The monkey has no idea what looks good and what doesn't, so it will consistently be right about 20-25% of the time.
2. Test takers will eliminate answer choices from the guessing pool based on a hunch or intuition. Simple but correct answers often get excluded, leaving a 0% chance of being correct. The monkey has no clue, and often gets lucky with the best choice.

This is why the process of elimination endorsed by most test courses is flawed and detrimental to your performance. Test takers don't guess; they make an ignorant stab in the dark that is usually worse than random.

$5 Challenge

Let me introduce one of the most valuable ideas of this course—the $5 challenge:

You only mark your "best guess" if you are willing to bet $5 on it.
You only eliminate choices from guessing if you are willing to bet $5 on it.

Why $5? Five dollars is an amount of money that is small yet not insignificant, and can really add up fast (20 questions could cost you $100). Likewise, each answer choice on one question of the test will have a small impact on your overall score, but it can really add up to a lot of points in the end.

The process of elimination IS valuable. The following shows your chance of guessing it right:

If you eliminate wrong answer choices until only this many remain:	Chance of getting it correct:
1	100%
2	50%
3	33%
4	25%

However, if you accidentally eliminate the right answer or go on a hunch for an incorrect answer, your chances drop dramatically—to 0%. By guessing among all the answer choices, you are GUARANTEED to have a shot at the right answer.

That's why the $5 test is so valuable. If you give up the advantage and safety of a pure guess, it had better be worth the risk.

What we still haven't covered is how to be sure that whatever guess you make is truly random. Here's the easiest way:

Always pick the first answer choice among those remaining.

Such a technique means that you have decided, **before you see a single test question**, exactly how you are going to guess, and since the order of choices tells you nothing about which one is correct, this guessing technique is perfectly random.

This section is not meant to scare you away from making educated guesses or eliminating choices; you just need to define when a choice is worth eliminating. The $5 test, along with a pre-defined random guessing strategy, is the best way to make sure you reap all of the benefits of guessing.

Secret Key #3 - Practice Smarter, Not Harder

Many test takers delay the test preparation process because they dread the awful amounts of practice time they think necessary to succeed on the test. We have refined an effective method that will take you only a fraction of the time.

There are a number of "obstacles" in the path to success. Among these are answering questions, finishing in time, and mastering test-taking strategies. All must be executed on the day of the test at peak performance, or your score will suffer. The test is a mental marathon that has a large impact on your future.

Just like a marathon runner, it is important to work your way up to the full challenge. So first you just worry about questions, and then time, and finally strategy:

Success Strategy

1. Find a good source for practice tests.
2. If you are willing to make a larger time investment, consider using more than one study guide. Often the different approaches of multiple authors will help you "get" difficult concepts.
3. Take a practice test with no time constraints, with all study helps, "open book." Take your time with questions and focus on applying strategies.
4. Take a practice test with time constraints, with all guides, "open book."
5. Take a final practice test without open material and with time limits.

If you have time to take more practice tests, just repeat step 5. By gradually exposing yourself to the full rigors of the test environment, you will condition your mind to the stress of test day and maximize your success.

Secret Key #4 - Prepare, Don't Procrastinate

Let me state an obvious fact: if you take the test three times, you will probably get three different scores. This is due to the way you feel on test day, the level of preparedness you have, and the version of the test you see. Despite the test writers' claims to the contrary, some versions of the test WILL be easier for you than others.

Since your future depends so much on your score, you should maximize your chances of success. In order to maximize the likelihood of success, you've got to prepare in advance. This means taking practice tests and spending time learning the information and test taking strategies you will need to succeed.

Never go take the actual test as a "practice" test, expecting that you can just take it again if you need to. Take all the practice tests you can on your own, but when you go to take the official test, be prepared, be focused, and do your best the first time!

Secret Key #5 - Test Yourself

Everyone knows that time is money. There is no need to spend too much of your time or too little of your time preparing for the test. You should only spend as much of your precious time preparing as is necessary for you to get the score you need.

Once you have taken a practice test under real conditions of time constraints, then you will know if you are ready for the test or not.

If you have scored extremely high the first time that you take the practice test, then there is not much point in spending countless hours studying. You are already there.

Benchmark your abilities by retaking practice tests and seeing how much you have improved. Once you consistently score high enough to guarantee success, then you are ready.

If you have scored well below where you need, then knuckle down and begin studying in earnest. Check your improvement regularly through the use of practice tests under real conditions. Above all, don't worry, panic, or give up. The key is perseverance!

Then, when you go to take the test, remain confident and remember how well you did on the practice tests. If you can score high enough on a practice test, then you can do the same on the real thing.

General Strategies

The most important thing you can do is to ignore your fears and jump into the test immediately. Do not be overwhelmed by any strange-sounding terms. You have to jump into the test like jumping into a pool—all at once is the easiest way.

Make Predictions

As you read and understand the question, try to guess what the answer will be. Remember that several of the answer choices are wrong, and once you begin reading them, your mind will immediately become cluttered with answer choices designed to throw you off. Your mind is typically the most focused immediately after you have read the question and digested its contents. If you can, try to predict what the correct answer will be. You may be surprised at what you can predict.

Quickly scan the choices and see if your prediction is in the listed answer choices. If it is, then you can be quite confident that you have the right answer. It still won't hurt to check the other answer choices, but most of the time, you've got it!

Answer the Question

It may seem obvious to only pick answer choices that answer the question, but the test writers can create some excellent answer choices that are wrong. Don't pick an answer just because it sounds right, or you believe it to be true. It MUST answer the question. Once you've made your selection, always go back and check it against the question and make sure that you didn't misread the question and that the answer choice does answer the question posed.

Benchmark

After you read the first answer choice, decide if you think it sounds correct or not. If it doesn't, move on to the next answer choice. If it does, mentally mark that answer choice. This doesn't mean that you've definitely selected it as your answer choice, it just means that it's the best you've seen thus far. Go ahead and read the next choice. If the next choice is worse than the one you've already selected, keep going to the next answer choice. If the next choice is better than the choice you've already selected, mentally mark the new answer choice as your best guess.

The first answer choice that you select becomes your standard. Every other answer choice must be benchmarked against that standard. That choice is correct until proven otherwise by another answer choice beating it out. Once you've decided that no other answer choice seems as good, do one final check to ensure that your answer choice answers the question posed.

Valid Information

Don't discount any of the information provided in the question. Every piece of information may be necessary to determine the correct answer. None of the information in the question is there to throw you off (while the answer choices will certainly have information to throw you off). If two seemingly unrelated topics are discussed, don't ignore either. You can be confident there is a relationship, or it wouldn't be included in the question, and you are probably going to have to determine what is that relationship to find the answer.

Avoid "Fact Traps"

Don't get distracted by a choice that is factually true. Your search is for the answer that answers the question. Stay focused and don't fall for an answer that is true but irrelevant. Always go back to the question and make sure you're choosing an answer that actually answers the question and is not just a true statement. An answer can be factually correct, but it MUST answer the question asked. Additionally, two answers can both be seemingly correct, so be sure to read all of the answer choices, and make sure that you get the one that BEST answers the question.

Milk the Question

Some of the questions may throw you completely off. They might deal with a subject you have not been exposed to, or one that you haven't reviewed in years. While your lack of knowledge about the subject will be a hindrance, the question itself can give you many clues that will help you find the correct answer. Read the question carefully and look for clues. Watch particularly for adjectives and nouns describing difficult terms or words that you don't recognize. Regardless of whether you completely understand a word or not, replacing it with a synonym, either provided or one you more familiar with, may help you to understand what the questions are asking. Rather than wracking your mind about specific detailed information concerning a difficult term or word, try to use mental substitutes that are easier to understand.

The Trap of Familiarity

Don't just choose a word because you recognize it. On difficult questions, you may not recognize a number of words in the answer choices. The test writers don't put "make-believe" words on the test, so don't think that just because you only recognize all the words in one answer choice that that answer choice must be correct. If you only recognize words in one answer choice, then focus on that one. Is it correct? Try your best to determine if it is correct. If it is, that's great. If not, eliminate it. Each word and answer choice you eliminate increases your chances of getting the question correct, even if you then have to guess among the unfamiliar choices.

Eliminate Answers

Eliminate choices as soon as you realize they are wrong. But be careful! Make sure you consider all of the possible answer choices. Just because one appears right, doesn't mean that the next one won't be even better! The test writers will usually put more than one good answer choice for every question, so read all of them. Don't worry if you are stuck between two that seem right. By getting down to just two remaining possible choices, your odds are now 50/50. Rather than wasting too much time, play the odds. You are guessing, but guessing wisely because you've been able to knock out some of the answer choices that you know are wrong. If you are eliminating choices and realize that the last answer choice you are left with is also obviously wrong, don't panic. Start over and consider each choice again. There may easily be something that you missed the first time and will realize on the second pass.

Tough Questions

If you are stumped on a problem or it appears too hard or too difficult, don't waste time. Move on! Remember though, if you can quickly check for obviously incorrect answer choices, your chances of guessing correctly are greatly improved. Before you completely give up, at least try to knock out a couple of possible answers. Eliminate what you can and then guess at the remaining answer choices before moving on.

Brainstorm

If you get stuck on a difficult question, spend a few seconds quickly brainstorming. Run through the complete list of possible answer choices. Look at each choice and ask yourself, "Could this answer the question satisfactorily?" Go through each answer choice and consider it independently of the others. By systematically going through all possibilities, you may find something that you would otherwise overlook. Remember though that when you get stuck, it's important to try to keep moving.

Read Carefully

Understand the problem. Read the question and answer choices carefully. Don't miss the question because you misread the terms. You have plenty of time to read each question thoroughly and make sure you understand what is being asked. Yet a happy medium must be attained, so don't waste too much time. You must read carefully, but efficiently.

Face Value

When in doubt, use common sense. Always accept the situation in the problem at face value. Don't read too much into it. These problems will not require you to make huge leaps of logic. The test writers aren't trying to throw you off with a cheap trick. If you have to go beyond creativity and make a leap of logic in order to have an answer choice answer the question, then you should look at the other answer choices. Don't overcomplicate the problem by creating theoretical relationships or explanations that will warp time or space. These are normal problems rooted in reality. It's just that the applicable relationship or explanation may not be readily apparent and you have to figure things out. Use your common sense to interpret anything that isn't clear.

Prefixes

If you're having trouble with a word in the question or answer choices, try dissecting it. Take advantage of every clue that the word might include. Prefixes and suffixes can be a huge help. Usually they allow you to determine a basic meaning. Pre- means before, post- means after, pro - is positive, de- is negative. From these prefixes and suffixes, you can get an idea of the general meaning of the word and try to put it into context. Beware though of any traps. Just because con- is the opposite of pro-, doesn't necessarily mean congress is the opposite of progress!

Hedge Phrases

Watch out for critical hedge phrases, led off with words such as "likely," "may," "can," "sometimes," "often," "almost," "mostly," "usually," "generally," "rarely," and "sometimes." Question writers insert these hedge phrases to cover every possibility. Often an answer choice will be wrong simply because it leaves no room for exception. Unless the situation calls for them, avoid answer choices that have definitive words like "exactly," and "always."

Switchback Words

Stay alert for "switchbacks." These are the words and phrases frequently used to alert you to shifts in thought. The most common switchback word is "but." Others include "although," "however," "nevertheless," "on the other hand," "even though," "while," "in spite of," "despite," and "regardless of."

New Information

Correct answer choices will rarely have completely new information included. Answer choices typically are straightforward reflections of the material asked about and will directly relate to the question. If a new piece of information is included in an answer choice that doesn't even seem to relate to the topic being asked about, then that answer choice is likely incorrect. All of the information needed to answer the question is usually provided for you in the question. You should not have to make guesses that are unsupported or choose answer choices that require unknown information that cannot be reasoned from what is given.

Time Management

On technical questions, don't get lost on the technical terms. Don't spend too much time on any one question. If you don't know what a term means, then odds are you aren't going to get much further since you don't have a dictionary. You should be able to immediately recognize whether or not you know a term. If you don't, work with the other clues that you have—the other answer choices and terms provided—but don't waste too much time trying to figure out a difficult term that you don't know.

Contextual Clues

Look for contextual clues. An answer can be right but not the correct answer. The contextual clues will help you find the answer that is most right and is correct. Understand the context in which a phrase or statement is made. This will help you make important distinctions.

Don't Panic

Panicking will not answer any questions for you; therefore, it isn't helpful. When you first see the question, if your mind goes blank, take a deep breath. Force yourself to mechanically go through the steps of solving the problem using the strategies you've learned.

Pace Yourself

Don't get clock fever. It's easy to be overwhelmed when you're looking at a page full of questions, your mind is full of random thoughts and feeling confused, and the clock is ticking down faster than you would like. Calm down and maintain the pace that you have set for yourself. As long as you are on track by monitoring your pace, you are guaranteed to have enough time for yourself. When you get to the last few minutes of the test, it may seem like you won't have enough time left, but if you only have as many questions as you should have left at that point, then you're right on track!

Answer Selection

The best way to pick an answer choice is to eliminate all of those that are wrong, until only one is left and confirm that is the correct answer. Sometimes though, an answer choice may immediately look right. Be careful! Take a second to make sure that the other choices are not equally obvious. Don't make a hasty mistake. There are only two times that you should stop before checking other answers. First is when you are positive that the answer choice you have selected is correct. Second is when time is almost out and you have to make a quick guess!

Check Your Work

Since you will probably not know every term listed and the answer to every question, it is important that you get credit for the ones that you do know. Don't miss any questions through careless mistakes. If at all possible, try to take a second to look back over your answer selection and make sure you've selected the correct answer choice and haven't made a costly careless mistake (such as marking an answer choice that you didn't mean to mark). The time it takes for this quick double check should more than pay for itself in caught mistakes.

Beware of Directly Quoted Answers

Sometimes an answer choice will repeat word for word a portion of the question or reference section. However, beware of such exact duplication. It may be a trap! More than likely, the correct choice will paraphrase or summarize a point, rather than being exactly the same wording.

Slang

Scientific sounding answers are better than slang ones. An answer choice that begins "To compare the outcomes…" is much more likely to be correct than one that begins "Because some people insisted…"

Extreme Statements

Avoid wild answers that throw out highly controversial ideas that are proclaimed as established fact. An answer choice that states the "process should used in certain situations, if…" is much more likely to be correct than one that states the "process should be discontinued completely." The first is a calm rational statement and doesn't even make a definitive, uncompromising stance, using a hedge word "if" to provide wiggle room, whereas the second choice is a radical idea and far more extreme.

Answer Choice Families

When you have two or more answer choices that are direct opposites or parallels, one of them is usually the correct answer. For instance, if one answer choice states "x increases" and another answer choice states "x decreases" or "y increases," then those two or three answer choices are very similar in construction and fall into the same family of answer choices. A family of answer choices consists of two or three answer choices, very similar in construction, but often with directly opposite meanings. Usually the correct answer choice will be in that family of answer choices. The "odd man out" or answer choice that doesn't seem to fit the parallel construction of the other answer choices is more likely to be incorrect.

Special Report: Additional Bonus Material

Due to our efforts to try to keep this book to a manageable length, we've created a link that will give you access to all of your additional bonus material.

Please visit http://www.mometrix.com/bonus948/afoqt to access the information.